The importance of sustainable development has been realized for at least 60 years, even though the vast majority of people erroneously think this concept originated with the Brundtland Commission report of 1987 on Our Common Future. In spite of at least six decades of existence, we only have some idea as to what is NOT sustainable development rather than what is. SpringerBriefs on Case Studies of Sustainable Development identify outstanding cases of truly successful sustainable development from different parts of the world and analyze enabling environments in depth to understand why they became so successful. The case studies will come from the works of public sector, private sector and/or civil society. These analyses could be used in other parts of the world with appropriate modifications to account for different prevailing conditions, as well as text books in universities for graduate courses on this topic. The series of short monographs focuses on case studies of sustainable development bridging between environmental responsibility, social cohesion, and economic efficiency. Featuring compact volumes of 50 to 125 pages (approx. 20,000—70,000 words), the series covers a wide range of content—from professional to academic—related to sustainable development. Members of the Editorial Advisory Board: Mark Kramer, Founder and Managing Director, FSG, Boston, MA, USA Bernard Yeung, Dean, NUS Business School, Singapore

More information about this series at http://www.springer.com/series/11889

Jessica M. Williams · Vivian Chu · Wai-Fung Lam ·
Winnie W. Y. Law

Revitalising Rural Communities

 Springer

Jessica M. Williams
Centre for Civil Society and Governance
University of Hong Kong
Hong Kong, Hong Kong

Vivian Chu
Centre for Civil Society and Governance
University of Hong Kong
Hong Kong, Hong Kong

Wai-Fung Lam
Centre for Civil Society and Governance
University of Hong Kong
Hong Kong, Hong Kong

Winnie W. Y. Law
Centre for Civil Society and Governance
University of Hong Kong
Hong Kong, Hong Kong

ISSN 2196-7830   ISSN 2196-7849 (electronic)
SpringerBriefs on Case Studies of Sustainable Development
ISBN 978-981-16-5823-5   ISBN 978-981-16-5824-2 (eBook)
https://doi.org/10.1007/978-981-16-5824-2

This Springer imprint is published by the registered company Springer Nature Singapore Pte Ltd.
The registered company address is: 152 Beach Road, #21-01/04 Gateway East, Singapore 189721, Singapore

Acknowledgments

The book is based upon the collective efforts and experiences of the members of the Centre for Civil Society and Governance at The University of Hong Kong in their endeavour to study and promote rural sustainability in Hong Kong. We hope it adequately documents and recognises the wisdom and contribution of their hard work. The authors would like to express their utmost gratitude to the Lai Chi Wo Programme team, including but not limited to: Dr. S. Liao, Dr. B. Hau, Ms. K. Chick, Ms. A. Yau, Mr. R. Leung, Ms. S. Yiu, Ms. V. Leung, Ms. J. Wan, Mr. L. Li and Ms. Y. Zhang, for their contribution to the Programme.

The authors would like to express their appreciation to HSBC for funding and supporting the Living Water and Community Revitalisation: An Agricultural-led Action, Engagement and Incubation Programme at Lai Chi Wo (2013–2017) and the HSBC Rural Sustainability Programme (since 2017) initiated by the Centre. The HSBC Climate Solutions Partnership also provided support in the scaling-up of the Programme to the regional level. Without this support, the Programme would never have materialised.

Last but not least, the Programme would not have succeeded without the support and enthusiastic involvement of the Lai Chi Wo villagers and members of different communities of interest. Their passion and dedication to rural revitalisation enabled many hurdles to be overcome and continues to sustain the Programme's efforts as well as Lai Chi Wo's culture and spirit.

Contents

Abbreviations

AFCD	Agriculture, Fisheries and Conservation Department of the Hong Kong SAR Government
AIRI	APAC Initiative for Regional Impact
APAC	Asia Pacific Accreditation Cooperation
CCO	Countryside Conservation Office
DBH	Diameter at breast height
DSD	Drainage Services Department of the Hong Kong SAR Government
HKCF	Hong Kong Countryside Foundation
HKSAR/SAR	Hong Kong Special Administrative Region
HKU	The University of Hong Kong
HSBC	Hong Kong and Shanghai Banking Corporation
LCW	Lai Chi Wo
NGO	Non-government organisation
Programme	The Sustainable Lai Chi Wo: Living Water & Community Revitalization—An Agricultural-led Action, Engagement and Incubation Programme at Lai Chi Wo and the HSBC Rural Sustainability Project
Programme team	PSL, HKU and partner organisations responsible for the development and implementation of the Programme
PSL	Policy for Sustainability Lab of the University of Hong Kong
SDG	Sustainable Development Goal
SES	Social-ecological system
The Academy	The Academy for Sustainable Communities run by the Centre for Civil Society and Governance at the University of Hong Kong
UNDP	United Nations Development Programme
UNESCO	United Nations

Chapter 1
Rural Sustainability: Challenges and Opportunities

Abstract Rural areas across the world have experienced outmigration and loss of economic viability due to processes such as globalisation and increased urbanisation. The loss of rural areas, which provide important resources on which wider societies depend on, creates immense risks for the long-term well-being of the world's growing urban populations. Their revitalisation, therefore, has benefits for both rural and urban areas. The interconnections between rural and urban areas, in this respect, are, however, often overlooked. By exploring the case of Lai Chi Wo, a village in Hong Kong on the outskirts of the urban community, it is demonstrated how these connections can be enhanced and better managed through the revitalisation of a rural area, bringing vibrancy back to a near abandoned village.

1.1 Introduction

In many developed countries and regions, rural areas have experienced a rapid demographic change and socio-economic transition through the process of rapid industrialisation and urbanisation. The migration of rural populations to urban areas has led to a dramatic decline in rural populations (Bjorna & Aarsaether, 2009; McGreevy, 2012; Stead, 2011; Walser & Anderlik, 2004). The depopulation of rural communities has been accompanied by a widespread loss of farmland to urban encroachment, abandonment of rural housing and the degradation of public infrastructure and other services (Bjorna & Aarsaether, 2009; Li et al., 2014; McGreevy, 2012).

Researchers and development practitioners increasingly recognise the potential of rural regions in providing the backdrop and resources for developing sustainable models as these provide the needs and well-being of growing population. As such, institutions at global and national levels have begun to pay attention to rural development. This book responds to the demand for a better understanding of the complexities of building sustainable rural communities, focusing particularly on revitalising rural resources in the urban context. It draws primarily upon knowledge and experience accumulated over approximately nine years of dedicated revitalisation efforts in the

© University of Hong Kong 2021

J. M. Williams et al., *Revitalising Rural Communities*, SpringerBriefs on Case Studies of Sustainable Development, https://doi.org/10.1007/978-981-16-5824-2_1

periphery of urban Hong Kong, analysing a wide range of strategies adopted in the quest for sustainability that is modelled on a village named Lai Chi Wo.[1]

1.1.1 The Lai Chi Wo Case Study: A Brief Overview

Lai Chi Wo (LCW) was once a thriving rural community that fed its few hundred villagers and surrounding smaller villages (Chick, 2017). Migration to urban areas and European countries since the 1950s had eventually led to the village being completely abandoned within a few decades. By the 1990s, the last villager bid farewell to a village that had lost much of its traditional culture and socio-ecological interactions due to a lack of active management.

With almost 400 years of history, LCW is one of the largest remaining traditional Hakka villages in Hong Kong. In addition to the cultural and historical values embedded in the village, it is also located in a rare ecological hotspot, evidenced by it being nestled in country, marine and geo-parks. Previously following traditional terrace farming practices, rows of paddy fields and freshwater wetlands on the hillsides and at the foot of the hills used to contribute to the rich ecosystem of this area in the North-eastern part of Hong Kong's New Territories.

In an effort to forge a sustainable path for villages in the rural urban interface, a revitalisation project[2] was initiated by the University of Hong Kong (HKU) in 2013, with the support of the Hong Kong and Shanghai Banking Corporation Ltd. (HSBC)'s Hongkong Bank Foundation. Recognising the potential value that could be recovered in LCW, revitalisation efforts were first concentrated on this village. The initial four-year project made significant progress in raising public awareness, rebuilding the community and livelihoods, leading to the return of Indigenous villagers and the introduction of new settlers. This allowed the next four-year project 'HSBC Rural Sustainability', starting in 2017, to expand to mobilising actions of the local community and community of interests for the benefit of the wider society. It also aims to incubate a set of viable models that could be applicable to nearby villages in Hong Kong and similar rural areas situated in the urban context in other parts of the world.

These revitalisation models have received recognition from the United Nations Development Programme (UNDP) and the community was a finalist for the Equator Prize in 2019 for its outstanding community efforts to tackle climate change and contribute to sustainable development through innovative and nature-based solutions. In 2020, the LCW Programme achieved Special Recognition for Sustainable Development in the UNESCO Asia–Pacific Awards for Cultural Heritage Conservation.

---

[1] https://ccsg.hku.hk/ruralsd/en/pages/about/introduction-to-lai-chi-wo/.

[2] Full project name: "Sustainable Lai Chi Wo: Living Water & Community Revitalization—An Agricultural-led Action, Engagement and Incubation Programme at Lai Chi Wo".

All over the world, discordant attempts have been made to revitalise rural areas. Experiences from numerous case studies have led to the emergence of several principles, such as taking a participatory approach and the importance of approaching revitalisation with a scope beyond economic development, for successful revitalisation. A broad spectrum of rural revitalisation strategies, guided by these principles, have been tested in this programme. The tracking of the entire process enables in-depth discussions of issues that arise at different stages of the revitalisation process, the ways in which they can be addressed and the lessons learnt through the experimentation of numerous strategies. All of these experiences and insights are invaluable in building knowledge and understandings that can be applied to the revitalisation of resources at the interface of urban and rural areas in other regions.

1.2 Managing Resources in Peri-Urban Areas and Revitalisation Efforts

Rural areas around the world have experienced profound changes over recent decades. Rapid urbanisation has led to rural decline, and some rural areas have increasingly become characterised by urban influences, resulting in areas known as the peri-urban interface (Simon et al., 2006). While significant research has focused on the acceleration of urbanisation and how urban processes encroach upon rural areas (for example, discussions about the urban footprint), the revitalisation of rural resources through cautious management of urban and rural interconnections has received scant attention. As rural populations continue to decline, management regimes for common resources collapse and/or undergo transformation (Singh & Narain, 2019). As traditional institutions fall apart, rural communities and resources become fragile and vulnerable to increasing pressures related to urban processes.

Rural revitalisation can be considered as a process to reverse rural decline and that focuses on the creation and stimulation of opportunities that will generate local rural income and jobs, while preserving and sustaining the dynamics and features characteristic of rural life (Kenyon, 2008; Meyer, 2014). Revitalisation can improve rural governance by ensuring the accountability of local governments in delivering a high standard of service (Steiner & Fan, 2019a). The positive outcomes that rural revitalisation aims to achieve include stabilising and increasing the local population, diversifying the economy and employment base, maintaining an acceptable level of service and preserving special rural attractions (Meyer, 2014). At the same time, traditional institutions around common resource use could be maintained, revived or transformed for improved management. As a result of rural revitalisation, rural areas become more productive, sustainable, healthy and attractive places to live (Steiner & Fan, 2019b).

1.2.1   Common Issues with Rural Revitalisation Efforts

A common response from governments and development organisations to combat rural decline is the introduction of policies that focus on diversifying the economy, stimulating enterprise and modernising infrastructure in rural areas (Bjorna & Aarsaether, 2009; Natsuda et al., 2012; van der Ploeg et al., 2000). These measures are often implemented through top-down policy, planning and investment initiatives and so yield mixed results (Abrams & Gosnell, 2012; Berke et al., 2013; Cabanillas et al., 2013; Li et al., 2014; Pasakarnis et al., 2013).

Throughout the last century, rural regions were primarily developed through top-down, government driven integrated development projects. Institutions such as the World Bank and USAID were often highly involved in rural development projects in the 1970s and 1980s (De Janvry et al., 2002). These projects were predicated on the basis that rural regions should be developed through expanding and modernising agricultural and primary production (Liu & Li, 2017). These top-down projects were criticised for being undemocratic and were seen as ineffective as they largely excluded local stakeholders and seldom considered the local resource pool in their decision-making and strategic planning (Liu & Li, 2017). Consequently, they often only achieved limited success, with project results' being unsustainable once state support ended (De Janvry et al., 2002).

Aligned with a top down approach to rural revitalisation, rural development traditionally was overly focused on agriculture, taking a narrow approach to farming and food security (Meyer, 2014). Traditional economic development perspectives saw agriculture as supportive to industry. Its role was to provide inputs for industrialisation, contribute to foreign exchange, provide rural communities with income and assist with capital investment (Meyer, 2014; Todaro & Smith, 2011). Therefore, a common solution for providing economic development to rural areas was often industrialisation.

It has come to be recognised that one of the requirements for successful rural revitalisation is the involvement of the local community in the planning and implementation process (De Satge, 2010; Meyer, 2014). Following, rural development has a strong social component, forming part of the UNDP's 'rural triple win model', alongside economic development and sustainable development (UNDP, 2012). As such, a top-down approach neglects the importance of local community involvement.

Conversely, the approach towards rural development that emerged in the 1990s took a bottom-up perspective. This employs a range of participatory methods (e.g. community cooperatives, interviews and surveys) to engage local communities and allows rural stakeholders to lead and facilitate development. Rather than focusing on agricultural production, bottom-up rural development recognises the importance of economic diversification and the opportunities for developing the services industry in rural areas through local initiatives and enterprises.

However, in many instances a solely bottom up approach to rural development can also be problematic. This is as there is often difficulty in coordinating and representing the diverse interests of the local people. In particular, the core groups that drive initiatives may not actually represent the interests of the wider community (McDonagh, 2001; Woods, 2011). There may also be issues of accountability and power as well as the extent that different sectors of the community actively participate in projects (Edwards et al., 2000, 2003; Storey, 1999; Woods, 2011). Shifting responsibility for rural development to local communities may also create an uneven geography for regeneration as some communities are better equipped to initiate projects or acquire funds (Edwards et al., 2000; Jones & Little, 2000; Woods, 2011). It may also not be possible to regenerate some rural communities once their traditional economic activities have declined or disappeared (Woods, 2011). As such, it is argued that a more collaborative approach that incorporates the needs of the wider community, linking rural and urban areas, is required.

1.2.2 Challenges and Opportunities of Revitalisation in the Peri-Urban Context

Rapid urbanisation has resulted in the loss of farmlands to industrial and residential uses as well as the loss of populations working with agriculture (Nilsson et al., 2013). Globally, capitalist agriculture facilitated by trade liberalisation has produced a model focused on high volume production in areas with low costs, with the products then being transported across the world to serve mass markets (Robinson, 2004). This has resulted in the breakdown of geographical dependence on local resources and a spatial remoteness within the global agricultural food system between production and consumption as well as the deterioration of local rural–urban relationships (Clark et al., 2013; McClintock, 2010). Consequently, agricultural activities decline and farms and farmland are abandoned due to marginalisation and inability to compete with corporate agribusinesses (Wästfelt & Zhang, 2016).

Arguments made by researcher on sustainable management of the peri-urban interface echo those discussed above in the context of rural revitalisation, such as the need for a holistic approach towards building livelihoods and natural resource management (McGregor et al., 2006) and addressing socio-economic needs of a diversity of stakeholders (Shaw et al., 2020). Certain characteristics of the peri-urban context, however, bring about unique challenges and opportunities to revitalisation efforts. Improved connections to the urban can be manifested in various ways for the rural area in the peri-urban interface, which includes the availability of public resources and support, improved infrastructure and accessibility, proximity to markets and immigration. Proximity to markets, skills and expertise in urban areas provides a much wider range of opportunities for livelihood diversification and the development of innovative models for communities in the rural–urban interface.

Being better connected to urban processes, however, does not lead to entirely positive outcomes. Increased availability of public resources could also mean increased government control. In the peri-urban, this could lead to conflicts between formal and informal governance (McGregor et al., 2006). In addition, although the relative availability of manpower to rural communities in the peri-urban interface is a great asset facilitating their revitalisation, increased human activity inevitably puts pressure on natural resources.

1.3 Hong Kong's Rural Landscape

Hong Kong, a Special Administrative Region (SAR) of China, is located in the Pearl River Delta Region of China's south coast. Due to Hong Kong's history, urbanisation began on Hong Kong Island and Kowloon peninsula, which were initially colonised by the British in the early eighteenth century. The British leased the New Territories in the late eighteenth century and the colonial government left the territory largely undisturbed. The New Territories was a primitive rural landscape, where large agricultural areas and thousands of villages could be found (Chiu & Hung, 1997). Prior to World War II, farmers undertook subsistence agriculture and livestock production, exported rice and imported other food from the Indo-China region (Grant, 1960; Nichols, 1976). An influx of refugees after the war saw vegetable self-sufficiency increase to 75% in 1961 (Airres, 2005).

However, from the 1960s, population pressure in urban Hong Kong meant that the New Territories began to undergo social and economic development, displacing agriculture (Strauch, 1984). In addition, in the 1970s and 1980s Hong Kong's economy started to move towards light industry. As a result, many village residents of the New Territories were evicted to make way for new towns, reservoirs and highways. This also led to a lack of public investment in agriculture, which coupled with a decline in farming resources contributed to the fall in local production (Tam, 2018). It was also around this time that residents started to abandon agricultural land as they moved into urban areas or emigrated (Chan, 1999).

From 1953 to 1975, abandoned agricultural and paddy land increased from 8.1% to 29% of total agricultural area (Airres, 2005). Land for agricultural production has been in steady decline due to urbanisation and industrialisation (Chan, 1999). Consequently, about 90% of Hong Kong's food is imported, with the majority coming from Mainland China (FHB, 2010). In 2015, 84.8% of total agricultural land was abandoned or fallow (excluding areas occupied by fish ponds) (AFCD, 2016). Currently, 3170 ha of land is zones as Agriculture (approximately 4.6% of Hong Kong), of this land, 1418 ha could be put to agricultural use but only 328 ha are under active cultivation, the rest is left idle or abandoned (Legco, 2016).

Abandoned agricultural land is a significant deterrent to agricultural activity in Hong Kong. There are currently about 2500 farms in Hong Kong, which employ about 4300 people (AFCD, 2020a, 2020b). The average farm size in the New Territories is only 0.2 ha, reflecting the small scale subsistence type agriculture of the past (Plan, 2016). Of this land, most of it is characterised by fragmented private ownership and so must be rented from landowners (Ko, 2013). Landowners often only provide short term leases (Yau, 2014). There is increasing competition of land use for development or storage purposes, which leaves little incentives for landowners to rent their land for agricultural purposes. Many landowners are willing to leave their land idle and wait for a high development offer (Yau, 2014). In addition, since the implementation of the construction waste disposal charging scheme in 2006, illegal dumping of construction waste on agricultural land has been a common problem leading to land degradation (Alderson, 2018; WWF HK, 2016).

Government attempts to revitalise agriculture in Hong Kong appear to have fallen short. The government implemented the 'Agricultural Land Rehabilitation Scheme' in 1988 to assist farmers in renting private land. From 2008 to 2018, 484 applications were submitted but only 165 of these were successful. A 'New Agricultural Policy' was introduced in 2016, which established a HK $500 million 'Sustainable Agricultural Development Fund' to aid projects that enhance agricultural productivity and output or to enable farmers to move to sustainable or high value added operations. By the end of 2017, 20 general applications were received but none had been approved (Bai et al., 2018).

The Fund's application process has been criticised for being too complicated for farmers and that the vetting committee lacks agricultural practitioners. There is also scepticism about the policy itself, in particular its flagship programme, the Agricultural Park, on whether it truly benefits ordinary farmers. The government intends for the park to nurture agro-businesses and agro-technology to operate on a commercial scale. However, farmers believe the park is a public relations attempt by the government and does not actually address the needs of the farmers. The programme has been criticised for not supporting traditional farming practices and for focusing solely on intensive, commercial and profit orientated agricultural practices.

As such, these government policies and measures fall short in providing practical help for local individual farmers and those who want to enter the agricultural sector. The 'New Agricultural Policy' conceptualised farming as a technological intensive business, which goes against the culture and heritage of farming in Hong Kong and so is likely to be unsustainable and impractical (Bai et al., 2018).

Much of the undeveloped parts of the New Territories remains protected under the Country Parks Ordinance, which was brought into force in 1976 (amended in 1995). The Ordinance provides a legal framework for the designation, development and management of country parks and special areas in Hong Kong. Under this ordinance, 24 country parks have been designated for nature conservation, outdoor education and recreation as well as 22 areas designated areas of nature conservation (AFCD, 2020a, 2020b). The total area of Hong Kong's Country Parks makes up 443 km^2 out of 1108 km^2 of total land area for the SAR (GovHK, 2020). Villages and agricultural land have mostly been excluded from the boundaries of country parks, creating remote pockets of land known as enclaves, which limited the villages' accessibility and development options (Fig. 1.1).

In 2004, the New Nature Conservation Policy was introduced to regulate, protect and manage natural resources that are considered important for sustainably conserving biological diversity, while accounting for social and economic considerations. This includes the identification and management of areas designated as protected for the conservation of biological diversity, which includes ecosystems

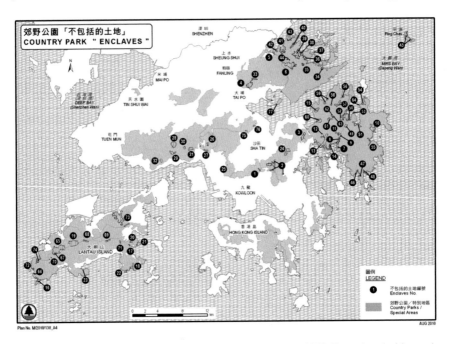

Fig. 1.1 Map showing all 77 Country Park Enclaves in Hong Kong 2010 (Reproduced with permission from the Agriculture, Fisheries and Conservation Department of the Government of Hong Kong, EPD)

and important habitats. It also covers the rehabilitation of degraded ecosystems and, where practical, the recovery of threatened species. The policy extends to include sites under private ownership and has identified 12 priority sites for enhanced conservation. To this end, the Nature Conservation Management Agreement and the Public–Private Partnership scheme were introduced (EPD, 2019).

The Management Agreement provides funding to enable NGOs to enter into management agreements with landowners of priority sites to enhance conservation. The scheme was extended in 2011 to cover country park enclaves and private land within country parks, currently there are seven such Management Agreement projects, including one at LCW, to rehabilitate and conserve these areas. The Public–Private Partnership scheme allows developments of an agreed nature at ecologically and less sensitive parts of the priority sites, as long as the developer conserves and manages the rest of the site. The developer is required to pledge a lump sum to support the conservation programme to the Environmental and Conservation Fund and identify a competent body to manage the ecologically sensitive portion of the site (EPD, 2019). The Public–Private Partnership scheme and Management Agreement were pioneering at the time due to their intention to engage landowners, protect ecologically sensitive private land and build partnerships between farm operators and NGOs. They have only achieved limited success, however, as there are criticisms from the environmental groups involved in the Management Agreements that the funding and time frame for the projects are insufficient, most projects are awarded on a two to three year basis, and so achieve little conservation results. Additionally, to date, there have been difficulties for those seeking Public–Private Partnership agreements to obtain approval (Ip & Hui, 2021).

1.4 The Lai Chi Wo Story

The story of Lai Chi Wo serves as an illustration of the fate of rural communities in Hong Kong. LCW has a long history as a Hakka[3] farming village as it was established in the seventeenth century. Located in the north-eastern part of Hong Kong's New Territories, the area sits between mountains and the sea and houses a diverse habitat of mature and secondary forests, shrub lands, agricultural wetlands, mangroves, mud flats, rocky shore and both freshwater and saltwater ecosystems. Villagers of LCW have increased their agricultural productivity by cutting the previously sloping landscape into a series of receding terrace steps, supported by stone, which is a traditional Hakka farming method for hilly terrain. The woodland to the back of LCW has been protected and managed by the villagers as Fung Shui Wood. Fung Shui Wood is characteristic of traditional South China farming villages and serves to protect the village from landslides and hill fires as well as moderates the microclimate and provides natural resources. It is believed that the large mangrove forest, found in the coastal area of LCW, was protected by the early villagers to defend the village against the tide, waves and storms (Fig. 1.2).

Despite the desertion of the village since the 1970s and 80s, the villagers have been able to maintain the continuity and coherence of their Indigenous culture and identity. A large proportion of those abroad still regularly visit the village. The villager's cultural bonds and traditions as well as the village's basic physical infrastructure remain intact. Many of the villagers have been open-minded about collaboration and partnership with people from outside, while at the same time adamant in upholding their stewardship of the natural and cultural heritage of their village.

In the early 2010s, Indigenous villagers of LCW reached out to HKU and environmental non-profit organisations in a bid to revitalise their village. The result was a collaborative Programme, which encompasses the revitalisation of agricultural practices, ecosystem monitoring, capacity building and villager engagement. It has also produced a range of socio-economic models and activities, some of which have gradually become self-sustained, and are now managed independently by individual villagers and other non-profit organisations and social ventures. The activities and initiatives of the Programme have inspired and attracted new ideas and possibilities

---

[3] In terms of culture, Hakka is a branch of 'Han' Chinese, who originated from central China. The Hakka people migrated south due to social unrest, wars or famines and as such, they were given the name 'Hakka' meaning 'travellers' or 'guests'. LCW village was founded by two Hakka clans, the Tsang's and the Wong's.

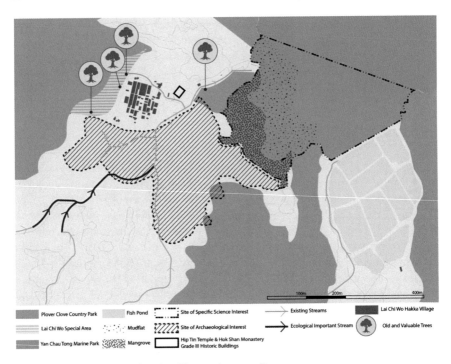

Legend:
- Plover Clove Country Park
- Lai Chi Wo Special Area
- Yan Chau Tong Marine Park
- Fish Pond
- Mudflat
- Mangrove
- Site of Specific Science Interest
- Site of Archaeological Interest
- Hip Tin Temple & Hok Shan Monastery Grade III Historic Buildings
- Existing Streams
- Ecological Important Stream
- Lai Chi Wo Hakka Village
- Old and Valuable Trees

Fig. 1.2 Map of LCW showing the village and surrounding area

from other interested individuals and groups. In turn, this has fostered the formation of new partnerships and collaborative efforts; an example is the development of locally produced food products.

The Programme adopts a collaborative approach in re-building and enriching the socio-economic, cultural and ecological capitals of LCW as well as to strengthen rural–urban linkages. Through the revitalisation process, flows of resources between the rural and urban areas have been established. This has led to significant changes to the village and, as a result, it is becoming increasingly characterised as peri-urban.

The Programme is developed and sustained through close partnerships between villagers and different stakeholder groups. Village affairs, which have been revitalised as a result of the Programme, remain governed by the Indigenous villagers and are supported by different stakeholder groups. The villagers of LCW have designed their own rule governing activities that impact the well-being of the village and organised collective actions to address problems. Such activities improved the local governance structures and processes and also allowed the creation of a productive relationships between the Indigenous villagers, new settlers and stakeholders.

A landscape approach is taken to achieve sustainable farmland revitalisation. This was done through the introduction of small scale eco-agriculture and high-yielding agroforestry, aided by low carbon technologies such as biochar and solar powered farming facilities. As a result of these practices, the Programme has restored six

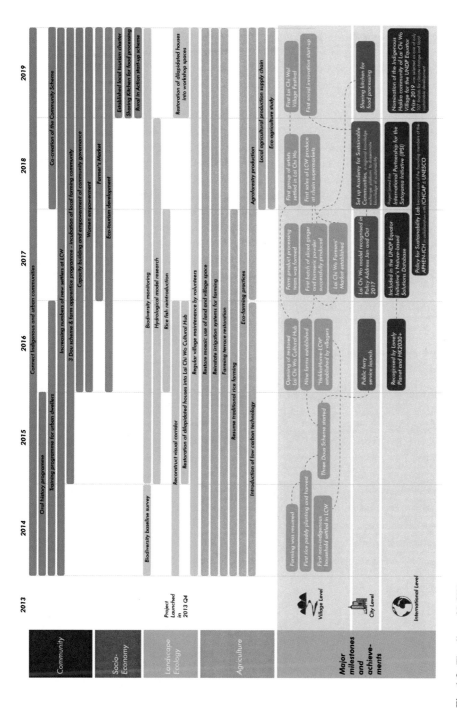

Fig. 1.3 Timeline of LCW programme

hectares of agricultural wetlands and open farmlands and, biodiversity of the area has been enhanced and monitored.

The Programme has also incubated and contributed to launching ten rural start-ups, which generate income from eco-farm produce and promote sustainable consumption and production. Such actions have allowed the village community to build its social resilience and develop collaborations for dealing with future challenges.

The revival of LCW is a successful model of community-led rural revitalisation in an Asian metropolis. This is not to say that this experience is only informative to the Asian context. This revitalisation model is specifically built given the challenges and opportunities found in rural communities within the urban context and, therefore, could be applicable to similar areas in any part of the world. Revitalisation in the peri-urban interface must make conscious efforts to anticipate and manage the impact of increased urban influences, while at the same time pursue opportunities that become available by being better connected with urban processes. Through re-establishing environmental-socio-economic connections between urban and rural areas, this case study demonstrates the importance of maintaining resources embedded in the peri-urban interface in building a sustainable and climate resilient city, for example in improving the city's food security, conserving biodiversity and enhancing an alternative low-carbon way of life (Fig. 1.3; Table 1.1).

Table 1.1 Actors involved in the LCW revitalisation process

Actor	Abbreviation	Brief description of role
Policy for Sustainability Lab	PSL	Programme team
The University of Hong Kong	HKU	Programme team
Hong Kong Countryside Foundation	HKCF	Partner organisation
Conservancy Association		Partner organisation
Project Green Foundation		Partner organisation
Agriculture, fisheries and conservation department	AFCD	HKSAR government department
Home Affairs Department		HKSAR government department
Drainage Services Department	DSD	HKSAR government department
Countryside Conservation Office	CCO	Government funding body
The Hong Kong and Shanghai Banking Corporation	HSBC	Funding body
Indigenous villagers		Living in Hong Kong and abroad—land and house owners
Pui Shing Tong		Village management committee
New settlers		From urban Hong Kong, many take up farming roles

1.5   Structure of the Book

In the next part of this book, the theoretical basis of collective governance is outlined, which underpins how sustainability is understood. This includes the importance of understanding social-ecological systems when governing natural resources and building of sustainable rural communities based on the concept of rural resilience and collaborative governance for peri-urban areas. The identified features of rural resilience are then explored in relation to the LCW case. The final chapter looks to the future, identifying enabling factors for successful sustainable revitalisation in such areas and how these can be scaled up and replicated.

References

Abrams, J. B., & Gosnell, H. (2012). The politics of marginality in Wallowa County, Oregon: Contesting the production of landscape consumption. *Journal of Rural Studies, 28*(1), 30–37.

AFCD. (2016). Departmental Annual Report 2015–2016. http://www.afcd.gov.hk/misc/download/annualreport2016/en/appendices_02.html. Accessed June 25, 2021.

AFCD. (2020a). Agriculture in HK. https://www.afcd.gov.hk/english/agriculture/agr_hk/agr_hk.html. Accessed June 25, 2021.

AFCD. (2020b). Hong Kong: The facts—country parks and special areas, the government of Hong Kong. https://www.afcd.gov.hk/english/country/cou_lea/the_facts.html. Accessed June 25, 2021.

Airriess, C. (2005). Governmentality and power in politically contested space: Refugee farming in Hong Kong's New Territories, 1945–1970. *Journal of Historical Geography., 31*, 763–783.

Alderson, T. A. (2018). *Supporting Hong Kong's local food economy: Developing a local food system assessment framework for intermediate markets.* HKU Thesis

Bai, F., Wan, V., & To, B. (2018). Back on the farm: The future of agriculture in Hong Kong. Hong Kong Free Press. https://hongkongfp.com/2018/04/01/back-farm-future-agriculture-hong-kong/. Accessed June 25, 2021.

Berke, P., Spurlock, D., Hess, G., & Band, L. (2013). Local comprehensive plan quality and regional ecosystem protection: The case of the Jordan lake watershed, North Carolina, USA. *Land Use Policy, 31*, 450–459.

Bjorna, H., & Aarsaether, N. (2009). Combaring depopulation in the northern periphery: Local leadership strategies in two Norwegian municipalities. *Local Governance Studies, 35*(2), 213–233.

Cabanillas, F. J. J., Aliseda, J. M., Gallego, J. A. G., & Jeong, J. S. (2013). Comparison of regional planning strategies: Countywide general plans in USA and territorial plans in Spain. *Land Use Policy, 30*(1), 758–773.

Chan, S. C. (1999). Colonial policy in a borrowed place and time: Invented tradition in the new territories of Hong Kong.*European Planning Studies, 9*(2).

Chick, H. L. (2017). The significance, conservation potential and challenges of a traditional farming landscape in an Asian metropolis: A case study of Lai Chi Wo, Hong Kong. *Journal of World Heritage Studies* (Special Issue), 17–23.

Chiu, S. W., & Hung, H. (1997). *The colonial state and rural protests in Hong Kong: Hong Kong Institute of Asia-Pacific Studies.* Chinese University of Hong Kong.

Clark, J. K., Munroe, D. K., & Ramsey, D. (2013). The relational geography of peri-urban farmer adaptation. *Journal of Rural Community Development, 8*(3), 15–28.

De Janvry, A., Sadoulet, E., & Murgai. R. (2002). Chapter 31 Rural development and rural policy. *Handbook of Agricultural Economics, 2*(A), 1593–1658.

De Satge, R. (2010). *Rural development in South Africa.* Phuhlisani Publishers.

Edwards, B., Goodwin, M., Pemberton, S., & Woods, M. (2000). *Partnership working in rural regeneration.* Policy Press.

Edwards, B., Goodwin, M., & Woods, M. (2003). Citizenship, community and participation in small towns: A case study of regeneration partnerships. In R. Imrie & M. Raco (Eds.), *Urban renaissance: New labour, community and urban policy* (pp. 181–204). Policy Press.

EPD (The Hong Kong Environmental Protection Department). (2010). Country Park Enclave. The Hong Kong Government. https://www.legco.hk/yr09-10/english/panels/ea/papers/devea0 728cb1-2721-1-e.pdf Accessed June 25, 2021.

EPD. (2019). New Nature Conservation Policy, the government of Hong Kong. https://www.epd. gov.hk/epd/english/environmentinhk/conservation/new_policy.html. Accessed June 25, 2021.

FHB. (2010). Frequently asked questions on food supply of Hong Kong. The Hong Kong Government. https://www.fhb.gov.hk/download/press_and_publications/otherinfo/110318_ food_supply_faq/e_food_supply_faq.pdf. Accessed June 25, 2021.

GovHK. (2020). Hong Kong: The facts country parks and conservation. The Hong Kong Government. https://www.gov.hk/en/about/abouthk/factsheets/docs/country_parks.pdf. Accessed June 25, 2021.

Grant, C. (1960). *The soils and agriculture of Hong Kong. Hong Kong: Government.* Government Press.

Ip, R., & Hui, J. (2021). *Striking a balance between development and conservation.* Our Hong Kong Foundation. https://www.ourhkfoundation.org.hk/en/report/18/land/striking-balance-bet ween-development-and-conservation. Accessed June 25, 2021.

Jones, O., & Little, J. (2000). Rural challenge(s): Partnership and new rural governance. *Journal of Rural Studies, 16,* 171–183.

Kenyon, P. (2008). *Rural revitalization and the need to create sustainable, healthy and resilient communities.* Bank of ideas. www.bankofideas.com.au. Accessed June 23, 2021.

Ko, W. (2013). LCQ10: Local farming industry development. http://www.info.gov.hk/gia/general/ 201306/26/P201306260312.htm. Accessed November 25, 2013.

Legco. (2016). LC Paper No. CB (2)767/15–16(03). http://www.legco.gov.hk/yr15-16/english/pan els/fseh/papers/fseh20160202cb2-767-3-e.pdf. Accessed June 25, 2021.

Li, Y. R., Liu, Y. S., Long, H. L., & Cui, W. G. (2014). Community-based rural residential land consolidation and allocation can help to revitalize hollowed villages in traditional agricultural areas of China: Evidence from Dancheng County, Henan Province. *Land Use Policy, 39,* 188–198.

Liu, Y., & Li, Y. (2017). Revitalize the world's countryside. *Nature News, 548*(7667), 275.

McClintock, N. (2010). Why farm the city? Theorizing urban agriculture through a lens of metabolic rift. *Cambridge Journal of Regional Economics and Society, 3,* 191–207.

McDonagh, J. (2001). *Renegotiating rural development in Ireland.* Aldershot.

McGreevy, S. R. (2012). Lost in translation: Incomer organic farmers, local knowledge, and the revitalization of upland Japanese hamlets. *Agriculture and Human Values, 29*(3), 393–412.

McGregor, D., Simon, D., & Thompson, D. (2006). The peri-urban interface in developing areas: The research agenda. In: D. McGregor, D. Simon, D. Thompson (Eds.), *The peri-urban interface: Approaches to sustainable natural and human resource use.* Earthscan.

Meyer, D. F. (2014). Exploration of solutions for revitalisation of rural areas in South Africa. *Mediterranean Journal of Social Sciences, 5*(4), 613.

Natsuda, K., Igusa, K. Wiboonpongse, A., & Thoburn, J. (2012). One village one product-rural development strategy in Asia: The case of OTOP in Thailand.

Nichols, E. (1976). The development of agriculture and fisheries in Hong Kong, Post 1946. In: AFCD (Ed.), *Agriculture Hong Kong,* 1st edn. Government Printer.

Nilsson, K., Pauleit, S., Bell, S., Aalbers, C., & Nielsen, T. A. S. (2013). *Peri-urban futures: Scenarios and models for land use change in Europe.* Springer Berlin Heidelberg.

Pasakarnis, G., Morley, D., & Maliene, V. (2013). Rural development and challenges establishing sustainable land use in Eastern European countries. *Land Use Policy, 30*(1), 703–710.

Plan, D. (2016). Planning for agricultural uses in Hong Kong. http://www.hk2030plus.hk/document/Planning%20for%20Agricultural%20Uses%20in%20Hong%20Kong_Eng.pdf. Accessed April 14, 2017.

Robinson, G. M. (2004). *Geographies of agriculture: Globalisation, restructuring and sustainability*. Pearson Education Limited.

Shaw, B. J., van Vliet, J., & Verburg, P. H. (2020). The peri-urbanization of Europe: A systematic review of a multifaceted process. *Landscape and Urban Planning, 196*, 103733.

Simon, D., McGregor, D., & Thompson, D. (2006). Contemporary perspectives on the peri-urban zones of cities in developing countries. In D. McGregor, D. Simon, & D. Thompson (Eds.), *The peri-urban interface: Approaches to sustainable natural and human resource use*. Earthscan.

Singh, A. K., & Narain, V. (2019). Fluid institutions: Commons in transition in the peri-urban interface. *Society and Natural Resources, 32*(5), 606–615.

Stead, D. R. (2011). Economic change in South West Ireland 1960–2009. *Rural History, 22*(1), 115–146.

Steiner, A., & Fan, S. (2019a). *A global rural crisis: Rural revitalization is the solution*. IFPRI blog: Issue post. https://www.ifpri.org/blog/global-rural-crisis-rural-revitalization-solution. Accessed June 24, 2021.

Steiner, A., & Fan, S. (2019b). *Global food policy report 2019*. IFPRI. http://ebrary.ifpri.org/utils/getfile/collection/p15738coll2/id/133129/filename/133348.pdf Accessed June 24, 2021.

Storey, D. (1999). Issues of integration, participation and empowerment in rural development: The case of LEADER in the Republic of Ireland. *Journal of Rural Studies, 15*, 307–315.

Strauch, J. (1984). Middle peasants and market gardeners: The social context of the 'vegetable revolution' in a small agricultural community in the New Territories, Hong Kong. In D. Faure, J. Hayes, & A. Birch (Eds.), *From village to city: Studies in the traditional roots of Hong Kong Society* (1st ed., pp. 191–205). Center of Asian Studies.

Tam, D. (2018). Food supply: Where food comes from. HK Foodworks. https://hkfoodworks.com/research/food-supply/. Accessed June 25, 2021.

Todaro, M. P., & Smith, S. C. (2011). *Economic development* (11th ed.). Pearson Education.

UNDP. (2012). *Triple wins for sustainable development: case studies for sustainable development in practice*. United Nations Development Programme.

van der Ploeg, J. D., Renting, H., Brunori, G., Knickel, K., Mannion, J., Marsden, T., de Roest, K., Sevilla-Guzmán, E., & Ventura, F. (2000). Rural development: From practices and policies towards theory. *Sociologia Ruralis, 40*(4), 391–408.

Walser, J., & Anderlik, J. (2004). The future of banking in America-rural depopulation: What does it mean for the future economic health of rural areas and the community banks that support them? *Banking and Financial Institution, 16*(3), 57–95.

Wästfelt, A., & Zhang, Q. (2016). Reclaiming localisation for revitalising agriculture: A case study of peri-urban agricultural change in Gothenburg, Sweden. *Journal of Rural Studies, 47*(A), 172–185.

Woods, M. (2011). *Rural geography: Processes, Responses and experiences in rural restructuring*. Sage Publications.

WWF-Hong Kong. (2016). LC Paper No. CB(1)331/16-17(01). http://www.legco.gov.hk/yr16-17/english/panels/ea/papers/ea20161219cb1-331-1-e.pdf Accessed June 25, 2021.

Yau, E. (2014). Call for incentives to help farmers stay on the land and boost Hong Kong's self-sufficiency. *South China Morning Post*. http://www.scmp.com/lifestyle/article/1555617/call-incentives-help-farmers-stay-land-and-boost-hong-kongs-self-suffiency. Accessed April 16, 2017.

Chapter 2
Rural Revitalisation: Building Rural Resilience Through Collaborative Governance

Abstract The sustainable management of an area necessitates understanding the complex interactions between human and social ecological systems. This perspective emphasises the relationship between humans and their environment and how communities can be determinative in sustaining healthy ecosystems. This supports that community involvement and participatory development are important pillars in rural revitalisation. Equally, this understanding draws attention to the impacts to the social-ecological system when conditions change, and the risks that such changes can hold for the delicate balance. Rural resilience, developed through collaborative governance processes, can be built into a social-ecological system to aid the system in dealing with unexpected shocks as well as to better adapt to the changing world. As such, the incorporation of rural resilience can contribute to building a sustainable rural community at the peri-urban interface.

2.1 Sustainability as Sustaining Social-Ecological Systems

The importance of community involvement and participatory development in rural revitalisation has been increasingly recognised since the 1970s (Li et al., 2014). More recently, there is growing recognition that the notion of humanity as decoupled from and in control of nature is a significant cause of societal vulnerability (Folke et al., 2002). Social vulnerability has been exacerbated by technological developments and economic activities based on a steady state view of environmental resources. This is particularly as the capacity of many ecosystems to generate resources and ecosystem services are vulnerable to external and internal change and, therefore, cannot be taken as certain (Folke et al., 2002, 2005).

Sustainable land and water management requires understanding of the interactions between humans and how they manage land and ecological systems (Baldwin et al., 2017; Fulton et al., 2011; Rammel et al., 2007). Striving for sustainability requires recognition of the coupled nature of social and ecological systems, which are inextricably linked (Baldwin et al., 2017). The complex connections that exist between people and nature is dynamic and non-linear, occurring between entities that are continuously changing and uncertain. As a result, synergies between economic development, technological change and the dynamic capacity of the natural resource base are required to support social and economic development (Folke et al., 2002). This requires moving away from conventional approaches that rest on assessments of maximum sustainable yields of individual species at a single broad scale to a more general focus on managing essential ecological processes that sustain the delivery of resources and ecosystem services across multiple scales (Folke et al., 2005).

Such an understanding has been termed as a 'social-ecological systems' (SES) approach (Berkes & Folke, 1992) and stresses the integrated concepts of humans within nature and that the separation of economic and social systems is contrived and arbitrary (Folke et al., 2005). Importantly, this approach recognises that ecosystem services are not generated solely by ecosystems but by SESs (Carpenter et al., 2005) and so people and their actions are essential to sustaining healthy, functioning ecosystems and their services.

A SES is composed of both bio-physical and social sub-systems, which form strongly coupled, complex and evolving integrated systems. The SES framework is often applied to the study of common pool resource management in the rural setting, for example in the study of irrigation systems (Lam, 2006; Lam & Chiu, 2016). Rural communities can be understood as a SES, which comprises natural and man-made resources, resource users, stakeholders and a governance structure. The survival of social systems depends on their interrelations with the system of natural resources and the environment, and natural resources are conditioned by the actions of the population (Schouten et al., 2009; Ambrosio-Albala & Delgado, 2008).

Understanding SESs is important for rural revitalisation efforts as the blind application of urban-oriented ideas, designs and technologies to rural areas is often unsuccessful. For example, the enthusiasm for private–public partnerships for developing rural areas as culturally unique townships and rural tourist resorts in China often results in attracting people more interested in exotic pleasure than the health of the ecosystem, wasting investment and fracturing local culture (Wang, 2020). Rural areas must be considered as complex SESs, which have the potential for sustainability, resilience and regeneration, in order for successful rural revitalisation (Wang, 2020).

2.1.1 *Rural Resilience of Social-Ecological Systems: Collaborative Rural Revitalisation*

The collapse of an SES is usually preceded by a loss of social, human, natural, physical and/or financial capital and the loss of resilience (Tenza et al., 2017). In general terms, ecological strain, particularly the depletion of natural resources and scarcity of human capital, such as economic stratification often result in complete collapse of a SES (Motesharrei et al., 2014). Some of these factors are endogenous, inherent in the SES itself and can include demographic changes, wealth distribution and natural resource use. Others are exogenous, being found outside the system and involve changes on a higher scale that impact internal dynamics, such as socioeconomic changes, natural disasters or technology changes (Tenza et al., 2017). The revitalisation of a rural area is essentially the revitalisation and transformation of a SES from one of near, or actual, collapse to a sustainable and resilient functioning system by addressing these exogenous and endogenous factors.

Accordingly, the LCW SES suffered from depopulation due to exogenous conditions brought about by economic changes, which resulted from urbanisation and shifts in food systems that reduced the viability of agriculture economically. This impacted resource use due to the loss of agricultural activities, which adversely impacted biodiversity, soil quality and increased the system's vulnerability to weather events such as severe rainstorms and typhoons. The revitalisation of the LCW SES, does not aim to simply recover the historic rural SES, as it has shown to not be viable in the modern world. Revitalisation in this case, therefore, required finding ways to ameliorate the outmigration of its population and develop new economic opportunities and resource management approaches, while preserving the cultural integrity of the village. As a result, it is essential to invest in capitals that will support such a change (Abel et al., 2006) and build resilience to prevent future collapse.

Resilience, as a key component under the SES framework, emerged in the 1980s (Carpenter et al., 2012a, 2012b) and has been developed for the rural context. The concept of 'rural resilience' was introduced by Heijman et al. (2007), which refers to the capacity of a rural region to adapt to changing external circumstances, in a way that maintains a satisfactory standard of living, while coping with its inherent ecosystem, economic and social vulnerability. Resilience determines the degree that a specific rural area is able to tolerate alteration before re-organising around a new set of structures and processes. Essentially, referring to how well a rural area can balance ecological, economic and social functions (Heijman et al., 2007; Schouten et al., 2009).

Resilience can be perceived as an approach for analysing, understanding and managing change in SESs or as a property of SESs (Folke et al., 2010). Resilience theory stresses that change is as normal a condition for SESs as stability. This is as the aim is to enable the system to respond to changing conditions in a manner that minimises losses to the system and its essential functioning (Redman, 2014). As such, central to resilience thinking are the concepts of adaptability, resilience and transformability (Folke et al., 2010).

Resilience theory provides the conceptual basis for sustainability (Carpenter et al., 2005) and so rural resilience provides insights into how to (re)develop and manage rural areas in a sustainable manner. The interconnected and co-evolving nature of agricultural and ecological systems in the production of food and maintenance of ecosystem functions and services makes resilience approaches appropriate to the rural setting (Scott, 2013). An underlying notion is one of balance, the idea that within a rural area, different balances are possible. In particular, it allows rural development policy to be reframed to include new modes of policy making based on adaptive collaboration networks. The environment becomes embedded into rural development through balancing the economy and ecosystems (Marsden, 2010; Scott, 2013).

Central to the resilience of a rural system is its adaptive capacity, which is determined by the degree that a rural area can tolerate change before it reorganises around a new set of structures and processes. When faced with a disturbance, stakeholders can either choose to maintain productivity within the same system or alter the system, seeking a new balance between ecology and economics. In this manner, sustainability is achieved through a change in the system (Folke et al., 2002; Heijman et al., 2007). However, a highly adaptive system does not necessarily equate to enhanced resilience. This is because rural resilience is comprised of resilience in several aspects including ecological, economic and cultural. A highly adaptive system in one aspect can lead to a loss of adaptability, and so resilience, in other areas (Schouten et al., 2009).

2.1.2 Building a Resilient Rural Social-Ecological System in the Peri-Urban Context

The peri-urban context is characterised by dynamism due to influences and flows of resources between the rural and urban as well as the changing nature of these flows under contextual conditions (McGregor et al., 2006; Singh & Narain, 2020). Common pool resources, however, have often been depleted under stronger urban influences due to land and water use pressure and governments' tendency to be biased towards the city (Narain & Vij, 2016). Therefore, villages in peri-urban areas call for highly adaptive and inclusive institutions. This is in line with research on adaptive governance of SESs, which argue for polycentric institutional arrangements, which are nested quasi-autonomous decision-making units that operate at multiple scales (Folke et al., 2005). This approach brings local knowledge directly into decision making and is more finely tuned to the needs of resource users as well as being more aware of learning and adapting (Berkes, 2009).

While being nested with higher levels of governance may offer the peri-urban SES more resources and support from the state, there are certain risks involved. For example, traditional management regimes based on local practices, norms and rules may clash with government control where overlapping jurisdictions give rise to conflicts and mismanagement. Where the state leans towards urbanisation and technological solutions such connections would prevent the establishment of a resilient rural SES in the peri-urban context.

2.2 Collaborative Governance and Adaptive Co-management for Rural Resilience

A range of theoretical constructs have been applied when discussing rural revitalisation and development, these include participatory rural development, rural networks and multifunctional rural development paradigms (Botes & van Rensburg, 2000; Chambers, 1994; Murdoch, 2000; Olfert & Partridge, 2010; Ryser & Halseth, 2010; van der Ploeg & Marsden, 2008). These often refer to concepts such as institutional thickness, actor networks and social capital in understanding rural development (Binns & Nel, 2003; Bjorna & Aarsaether, 2009; Lowe et al., 1995; Marsden, 2010). Such studies indicate that for successful and enduring community development, public participation, social capital and institutional thickness are required (Li et al., 2012, 2013, 2014).

Focusing on collective means of governing rural communities, this book mainly draws upon collaborative governance within SESs and the role of resilience in developing sustainability. Based predominantly upon these frameworks and understandings, this chapter identifies and discusses conceptual components commonly featured amongst them to identify the core components that lead to sustainable development and management. The concept and understandings of rural resilience is utilised as a means to provide a sharper focus on the rural context in the quest for sustainable rural revitalisation.

Sustainability is more likely to be attained through the adoption of collective governance processes such as adaptive co-management and collaborative governing. Before discussing common factors or characteristics put forward by these theories as contributors to achieving sustainable governance and management, they will be briefly introduced.

2.2.1 Collaborative Governance and Management of Social-Ecological Systems

Collaborative governance pertains to shared, negotiated and deliberative decision-making. It is a particularly popular approach for collective resources or ecosystem-based management (Bingham, 2011; Bodin et al., 2017). Particularly as no single actor group is able to achieve the revitalisation of a rural SES without supportive collaboration with other actor groups (Li et al., 2019). The theoretical framework emerged as a response to failures in implementing policies and programmes at the community level (Fung & Wright, 2001). There are various ways to define collaborative governance. Some definitions emphasise the role of the government—defining it as governing arrangements where "public agencies directly engage non-state stakeholders in collective decision-making process" (Ansell & Gash, 2007: 544), claiming that the "locus of authority often remains with government" (Gollagher & Hartz-Karp, 2013: 2348).

However, this is not taken to be the case here, rather, a combination of Emerson and Nabatchi's (2015) and Bingham's (2011) definition are deemed more fitting. According to Emerson and Nabatchi (2015), initiatives in collaborative governance may not be state-initiated and government involvement should not be assumed. In the case study of rural revitalisation presented in this book, the government was not involved in the early stages of the process. When it later became involved, its involvement played a supplementary role. Nonetheless, the importance of government support in facilitating the further development of rural communities is recognised amongst stakeholders and demonstrated in this case study.

Major theoretical frameworks on cross-sector collaboration developed within the last decade or so have been reviewed by Bryson et al. (2015) who reveals some common components. At the outset, all frameworks identify the need to include general antecedent conditions, such as the institutional environment and available resources that can affect the collaboration process and its outcomes. After these initial conditions, drivers and linking mechanisms are usually discussed. The central components regarding the collaborative processes and structures are then elaborated. Due to the difficulties in studying the collaborative processes and structures separately, particular areas of intersections include leadership, governance, technology and capacity and competencies have been highlighted. Many frameworks have also remained vigilant of endemic conflicts and tensions in collaborative processes and structures, for example power imbalances, tensions between inclusivity versus efficiency and autonomy versus interdependence (Bryson et al., 2015).

The structure of collaborative governance related theoretical frameworks is useful when developing an appropriate framework for this book's case study and so it is also mirrored in the structure of this chapter. Following, common factors and concepts identified in these frameworks have been categorised into three sub-sections—preceding conditions, collaboration process and structures.

2.3 Collaborative Governance: Factors Enhancing Rural Resilience

These theories suggest some common factors that increase the likelihood of success in building rural resilience. These factors include the presence of social capital (before and during collaboration), effective learning, adaptation and leadership during the collaboration process, a dynamic bridging organisation and the nesting of the collaborative institutions with institutions of other levels and scale.

Rural resilience is a significant contributor of sustainable rural communities, it has even been argued that sustainable rural communities are characterised by rural resilience. This is in response to the notion of rural decline and associated with enhancing villagers' livelihoods through behavioural changes and adaptation to new circumstance. Resilient rural communities, therefore, have the capacity to prevent challenges of negative externalities and to adapt to exogenous changes in a way that maintains a satisfactory standard of living and re-establishes social-ecological system integrity. As such, resilient and sustainable rural development has been found to be achieved through rural livelihood diversification, particularly the creation of economic activities that can meet potential urban demand, local entrepreneurship to establish and expand these activities, market orientated institutions to support these activities as well as strong social capital (Li et al., 2019). The presence of these factors amongst effective management of natural resources, therefore, can be taken to indicate the establishment of a sustainable rural SES. As such, how the conditions for developing resilient, and so sustainable, SESs are achieved by the factors identified by collaborative governance and resilience theories will be explored in the following chapter.

2.3.1 Prerequisite for Collaborative Rural Development and Management

The theories highlight several antecedent conditions that enhance the effectiveness or success of collaborative governance in achieving sustainability, amongst which select conditions are particularly relevant to the rural context. Such conditions include the state of the SES's resources, presence of entrepreneurial culture, levels of service, civil society and history of collective action (Bock, 2012: 18). The availability of resources, the ease of monitoring the condition of the resource as well as the rates of change in the resource are all influential on the effectiveness of resource governance regimes. Also influential is the institutional environment, which includes the structures and knowledge that support inter-sectoral cooperation as well as providing favourable conditions for community involvement. Social factors include social capital and social networks as well as the ability to address public issues and resources users' support for self-governance also contribute to creating a more favourable environment for more collaborative approaches (Bryson et al., 2015; Dietz et al., 2003; Wamsler, 2017).

2.3.1.1 Social Capital

Social capital is a prominent prerequisite for building rural resilience and is recognised to be enhanced through the collaboration process. Social capital refers to networks that produce trust, reciprocity and cooperation to mobilise resources. Its importance for community revitalisation is gaining recognition (Li et al., 2014; Ryser & Halseth, 2010). Building trust and networks serve to "foster a sense of shared goals" concerning the actions being undertaken (Gerlak & Heikkila, 2011). Shared values and social norms further aid trust building and so increases social capital (Brunckhorst & Marshall, 2007). Simply devolving management rights does not automatically result in adaptive co-management or similar self-organised government systems. Such systems require social networks and a certain level of social capital, as this cements adaptive capacity and collaboration (Folke et al., 2005).

When managing SESs, some theorists take a 'capitals' approach to understanding resilience at the community level. This approach focuses on the three capitals, social, economic and environment. The stronger the capitals are developed, the stronger the community's resilience. It is rare for capitals to be equally developed and as they are interlinked, changes to one capital can influence the others. Therefore, the range of factors that shape community resilience are non-linear, complex, interlinked and cumulative (Salvia & Quaranta, 2017).

Similarly, social resilience and cultural resilience, form part of social capital and are important conditions for rural resilience (Adger, 2000; Heijman et al., 2007). In the rural context, groups or communities are dependent on ecosystems for their livelihoods and economic activities. Therefore, social resilience is the ability of these groups to adapt to external social, political or environment stress and shocks (Adger, 2000, 2003, 2006). Social resilience occurs at the community, rather than the individual, scale (Scott, 2013). Cultural resilience ensures a sufficient human capital in the region (Heijman et al., 2007).

The social factors involved in SESs include levels of trust, learning and communication pathways, cooperation, strength of networks, bonding and bridging capitals, and community cohesiveness. These factors mediate the relationship between the economic and environmental components of a SES. In general, social capital in rural areas is weakened by out-migration and population ageing as this restricts intergenerational communication and knowledge transfer between the generations (Salvia & Quaranta, 2017). In particular, in the rural context, social capital is significantly weakened when villages are deserted.

Structural social capital, the set of social structures that allows for interactions amongst individuals (Nardone et al., 2010), can include social organisations, roles, rules and networks (Warner, 1999). Within structural social capital, formal networks are recognised groups related to politics, religion, hobbies etc., while informal ones consist of, for example, family, friends, neighbours and colleagues (Harpham, 2008). It is recognised that rural populations have stronger social structures (De Silva et al., 2007) and it has been found that local communities develop non-hierarchical dense ties, with close spatial proximity resulting in increased interpersonal interactions and connections (Go et al., 2013). Another study highlighted that for villagers, the village bonding network of relatives was the most important structural social capital (Zhang et al., 2013). As such, when dealing with rural areas the importance of community networks and relationships between the villagers is likely to be an important source of social capital.

Similarly, the link between rural communities and the environment is important for social capital. The exclusion of Indigenous people and local communities in conservation and management processes has been linked to the loss of biodiversity. The is commonly found in rural areas in the peri-urban context as they are affected by urban expansion, where conflicts over land use coupled with power imbalance lead to the marginalisation of communities previously dependent upon that land and its resources (Narain & Vij, 2016; Vij et al., 2018). To avoid such injustice and loss of biodiversity, the participation of the community in resource management is essential. This also builds social capital, which contributes to cooperation, social equity and efficient governance of a SES (Titumir et al., 2020). Social ties and a certain level of trust has to be attained for community revitalisation to take place and as the village becomes revitalised, social capital is further constructed.

It is important to note, however, that social capital is not always conducive to building rural resilience. For example, corrupt governance schemes and networks are a form of social capital that manipulates power and public trust to divert physical capital, at the expense of the wider society. Gangs and Mafia also use social capital as the foundation of their organisational structures (Brondizio et al., 2009). In the rural context, while social capital can create tight knit village communities, it can also act as a basis for hostility towards any 'outsiders' or outside interventions.

2.3.2 Collaborative Processes

2.3.2.1 Learning and Adaptation

The ability to learn and adapt is an important feature of the theories. Adaptation and learning are key components in building resilient rural areas. This is as change is a constant in agricultural practices and rural landscapes, therefore, continual learning and adaptation is necessary when utilising resources and for nurturing innovative behaviours (Glover, 2011).

Specifically, a lack of local knowledge often contributes to the failure of well-intended efforts at revitalising rural areas (Wang, 2020). Generally, local knowledge includes an ecological aspect as for social-ecological regeneration to occur, the local particulars of the environment and human-earth relations needs to be understood (Berkes, 2018). Such place specific knowledge is only possessed by those living and caring about the specific area. At the core of local knowledge is also the need to recognise and respect local peoples' opinions and autonomous roles as separate from bureaucracy and outside expertise (Berry, 1990; Wang, 2020). Local knowledge about indigenous peoples' relationship with the environment, folklore, beliefs and local memories has been found to be essential in the process of respecting local peoples' autonomy, planning and community cultural regeneration (Wang, 2020).

Collective learning comprises the collective process, which involves the acquisition of knowledge through diverse actions, such as trial and error, assessing information and disseminating new knowledge. This results in collective products, which are new ideas, strategies, rules or policies (Gerlak & Heikkila, 2011). Social learning, which includes practical learning though iterative practice, evaluation and action modification is also a key feature of adaptive management approaches (Berkes, 2009). The learning process and the resulting knowledge are centred on improving the understanding of the dynamics of the SES. This includes the interactions between sub-systems and the ways in which the SES of interest interacts with other SESs. Under the adaptive management framework, detailed knowledge regarding individual parts of the system is considered to be less important (Folke et al., 2005).

Given the inherent uncertainty in ecosystems, adaptive management provides that "management involves a continual learning process that cannot conveniently be separated into functions like 'research' and ongoing 'regulatory activities,' and probably never converges to a state of blissful equilibrium involving full knowledge and optimum productivity" (Walters, 1986: 8–9; Walters & Holling, 1990; Garmestani & Benson, 2013). The adaptability of a SES is its capacity to learn, combine experience and knowledge, adjust its response to changes in external drivers and internal processes and continue developing within the current stability domain or basin of attraction (Folke et al., 2010).

This adaptability, along with transformability, are core features of resilience. Transformability refers to the capacity to transform the landscape into a fundamentally new system, when ecological, economic or social structures make the existing system unsustainable (Folke et al., 2010). Resilience involves a system's ability to build its capacity for learning and adaptation at both an individual and collective level (Salvia & Quaranta, 2017). As such, cultivating resilience can be considered a process of social learning that utilised human capacities and knowledge to reduce vulnerability and risk in unknowable and unexpected circumstances (Hudson, 2010). However, a major challenge is in building knowledge, incentives and learning capabilities of governance institutions and organisations, which allow adaptive management of rural ecosystems as well as the incorporation of actors in new and imaginative roles (Gunderson & Folke, 2005; Heijman et al., 2007).

While adaptability is an important part of building resilience, it is important that it is balanced with stability. Stability allows legitimacy and long term relationships within collaborative governance structures to be built (Provan & Kenis, 2008). Adaptability allows for rapid responses to be developed to meet the changing needs and demands of the SES, however, stability is required for developing consistent responses and efficient management of the system over time (Provan & Kenis, 2008). Ecosystems may also require a level of stability as where adaptation results in efforts to increase the efficiency of resource use, it can result in a loss of response diversity. By increasing adaptability in one place, the adaptability, and so resilience, of another place may be lost. Increasing the adaptability of a system to specific or regular shocks can in turn decrease its general resilience to unknown shocks (Walker et al., 2006). As such, while adaptation is generally considered a positive outcome of revitalisation and key component of resilience, care must be taken to ensure the SES does not become 'over adapted' to specific phenomenon or activities.

2.3.2.2 Social Innovation

Social innovation includes changes in social relations, political arrangements or governance processes that result in the improvement of a social system. It is thus intended to improve society through "the mobilisation-participation processes", where the outcomes of innovations expect to "lead to improvements in social relations, structures of governance, [and] greater collective empowerment" (Castro-Arc et al., 2019; Moulaert et al., 2013: 2). Social innovation enhances a system's connections between socio-political levels and spatial scales, with the potential to link bottom-up initiatives with those at higher spatial levels. Essentially, this can result in bottom linked systems of governance and so more inclusive, diverse and adaptive governance systems (Castro-Arc et al., 2019). In a rural context, social innovation refers to changes in the social fabric of rural societies that are necessary for their survival. This can include changes in social relations and the capacity to organise collective action (Bock, 2016).

Social innovation increases the capability of a SES to respond to changes and challenges as it identifies the factors that foster transformation. It emerges from the needs, challenges, resources and institutions of the SES and so it encourages proactive sustainable governance of that system. Social innovation is place specific, responding to the particular needs of the area, reflects the choices and decisions of those involved and seeks to improve social and ecological conditions of that area (Castro-Arc et al., 2019). As such, collaborative networks play an important role in producing social innovation. It is where sectors converge in these networks that new and improved approaches to solve social problems are born (Kallis et al., 2009). This is as various actors become involved in discussions on social-ecological problems, they develop solutions and become socially engaged in addressing problems (Parra, 2013).

Depopulation of rural areas tends to result in the loss of the most entrepreneurial people, so it is only the most resourceful rural areas that are able to develop social innovations. This has the potential to compound existing inequalities and further spatial disparities. To compensate for this, social innovation does not necessarily have to originate locally, it can include the uptake of novel solutions developed elsewhere. Collaboration across space is also important for social innovation. As such, actors that are able to bridge regional and international contexts can be particularly valuable and there is the need to acknowledge that local development does not necessarily have to fully originate internally. Therefore, rural social innovation requires networking and building relations across the borders of the rural locality being examined (Bock, 2016).

Rural social innovation requires citizen engagement and entrepreneurialism, it focuses on citizen issues, socio-economic fragility and problems from welfare state reform and austerity measures. These reflect the changing political economy of rural development, which often receives limited state support (Bock, 2016).

2.3.2.3 Leadership

Alongside the need for strong social capital, leaders play an important role in ensuring the sustainable functioning of collaborative governance mechanisms. Leadership can be considered as a process, as such it is a transactional event that occurs between leaders and other stakeholders. As a result, leadership is an interactive event and relationships are highly significant (Avant et al., 2013). Effective leadership can help shape shared values, create organisational culture, build trust and manage conflicts (Folke et al., 2005). Leadership, whether it be in a formal or informal capacity, individual or organisational, can also spark learning processes, bring together diverse interests, ensure the commitment of participants, create a learning culture or transparency with information sharing, facilitate communication and willingness to experiment (Gerlak & Heikkila, 2011; Salvia & Quaranta, 2017). In brief, leaders develop and enable a vision for management that includes local knowledge and information from social networks (Garmestani & Benson, 2013).

Collaboration in governance networks requires some form of leadership and studies show that individual agency can play a strong role in increasing the adaptive management and governance of a system, thus building community resilience (Folke et al., 2005; Olsson et al., 2006, 2014). In fact, in watershed partnership governance, leadership has been found to be the second most frequent factor for successful partnerships after adequate funding (Leach & Pelkey, 2001). However, strong leadership does not necessarily entail more conventional ideas of top-down, command and control style management (Holling & Meffe, 1996). Such governing tends to be ineffective when dealing with complex changing systems. A lack of leadership, on the other hand tends to be associated with inertia in addressing problems in SESs as well as ineffectual management outcomes (Evans et al., 2015). In participatory processes, having an overly dominant leader can result in disengagement by the other participants and so is often not desirable in collaborative decision making (Evans et al., 2015; Pahl-Wostl et al., 2007).

Thus, effective leadership is better achieved through the actions and strategies of a number of actors. These actors each act to aid the system to move through different stages of innovation and transformation (Westley et al., 2013). It has also been demonstrated that to achieve greater community resilience, leadership needs to come from within the community and be generated by community needs instead of being formally conferred on individuals outside of the community (Faulkner et al., 2018). In rural systems, leadership is an important part of building community resilience (Walker & Salt, 2012). It needs to be understood within a complex adaptive system to appreciate the inter-sectorial nature of distributive leadership in regional areas. Leadership has also found to be essential in building community capacity (Banyai, 2009; Madsen & O'Mullan, 2014; Normann, 2013).

Resilient communities are constructed through active citizenship and activities that strengthen social networks such as volunteering, the ability to work collaboratively and draw out the best in each other (Bourgon, 2010). These skills are important for developing and cultivating social connections and networks. As such, it is important that there is a shared understanding of leadership as a community responsibility (Madsen & O'Mullan, 2014).

Developing collaborative leadership within rural communities rather than the skills and capacities of an individual has been found to yield promising result. Such collaborative leadership is built upon a shared vision for the future, which is able to exerts more influence on policy, attract more funding and builds community capacity (Madsen & O'Mullan, 2014). When acting in a similar manner, external organisations have also been discovered to be effective leaders, which will be discussed in greater detail in the following subsection. Regardless of who the leaders are, without leadership, SESs can fall into a state of inertia and so become vulnerable to changing conditions (Folke et al., 2005).

2.3.3   Structures

2.3.3.1   Bridging Organisations

To work collaboratively, different actors are destined to experience, to varying extents, shifts in their perspectives, knowledge frames and understanding of relevant processes and solutions. Boundary organisations or bridging organisations facilitate such processes. These two types of 'organisations' are similar in that they both offer the functions of communication, translation and mediation of scientific knowledge in accordance with the local context for the purpose of policy and action. However, bridging organisations have a broader scope in two ways. Firstly, boundary organisations refer to institutionalised forums or platforms for knowledge exchange and integration (Kallis et al., 2009), while bridging organisation may refer to both cross-sectoral platforms and organisations that provide similar functions such as NGOs or municipal organisations (Berkes, 2009; Folke et al., 2005). Second, and more importantly, bridging organisation cover a wider scope of functions. Instead of focusing primarily on improving communication between scientists and stakeholders, bridging organisation also emphasise the capacity of the network of stakeholders to mobilise knowledge and social memory during times of need (Folke et al., 2005).

As interested parties are gathered and engaged in a shared learning and knowledge generation process through the bridging/boundary organisation, a new language is developed to further facilitate cross-sector communication and collaboration (Kallis et al., 2009). Having experienced knowledge exchange and perspective shifts, actors then engage in forging an agreement on certain issues or plans, while simultaneously developing a new set of vocabulary to facilitate further communication.

Bridging organisations, such as NGOs or research institutes, perform the roles of visionary leaders and catalysers for collaboration or fill gaps left by formal institutions and sometimes perform essential functions by crafting effective responses (Berkes, 2009; Folke et al., 2005; Olsson et al., 2007). These organisations provide facilitation, leadership and social incentives for collaboration, all of which are essential components for building the community's capacity to adapt to change (Folke et al., 2005; Hahn et al., 2006).

The ability of bridging organisations to deal with uncertainty and craft timely responses is built upon the extent to which they enable the mobilisation of knowledge, including the ease of recording and retrieval of accumulated knowledge and experiences (Folke et al., 2005; Olsson et al., 2007). Under the framework of adaptive co-management, bridging organisations have no optimal form and must continuously adapt and remain flexible. Its organisational structure could resemble an adhocracy, as needs and issues arise, ad hoc projects are developed, each mobilising actors from the social network (Hahn et al., 2006).

Bridging organisations play an essential role in linking sectors and geographic scales (Berkes, 2002). They are able to support civil society networks that enable the representation of potentially marginalised, rural populations in development decision making (Ratner & Allison, 2012). Bridging organisations in the rural context ensure that local values and cultures are not lost when implementing change (Leys & Vanclay, 2011). This is particularly important when local knowledge is based on a different epistemology and worldview than government science (Berkes, 2002). This can help facilitate a connection between social learning and policy at higher governance levels (de Kraker, 2017).

Bridging organisations have been found to play an essential role in natural resources governance as they build local institutions, provide horizontal and vertical linkages and increase public education and innovations. They are also able to mediate connections between actors that are otherwise unconnected, attract new knowledge and resources from outside the natural resources management system, which are required for social learning and transformation. In this way, they are able to increase the SES's adaptability and resilience (Stewart & Tyler, 2019; Walker & Salt, 2006, 2012).

2.3.3.2   Polycentric Institutional Arrangements

While some problems encountered in a SES can be resolved through the efforts of local community members this is, in many instances, not feasible. Most problems will involve more than one SES either of a similar scale or of a larger scale at a higher level. Immensely complex issues not only call for the collaboration of a wide range of actors but that the problem-solving efforts draw upon different levels of authority (Herzberg, 2020). Solutions need to be based upon the complementary efforts of all relevant institutions (Lam, 2006).

Other than to solve problems, a well-functioning self-governing system will almost always require recognition from a higher level of government. This has also been discovered as one of the key features of a robust common property or self-governing regime. A robust SES has the capacity to continue to meet a performance objective in the face of uncertainty and shocks (Janssen & Anderies, 2007). Other than the coupling of self-governing systems with authorities, a robust system relies upon the factors discussed above such as the development of a broad spectrum of solutions and social memory (Lam, 2006).

The importance of a nested and polycentric institutional arrangement is highlighted in the study of self-governing regimes. Community members have demonstrated both the ability to design self-governing rules and to draw on the appropriate level of authority corresponding to conflicts that they faced (Herzberg, 2020). In many ways, an effective self-governing regime is about striking the right balance between decentralised and centralised control, where autonomous decision-making institutions are nested amongst local and/or higher organisations in a polycentric institutional arrangement (Folke et al., 2005).

Such institutional arrangements have been found to be important in building or enabling the capacity to manage resilience. Polycentric arrangements create opportunities for locally appropriate institutions to evolve as monitoring and feedback loops are tightened and the associated institutional incentives are enhanced (Berkes & Folke, 1998). This allows local government arrangements to develop in a way that better suits the social and ecological contexts and dynamics of specific locations. Local knowledge is able to inform local actions in a way that a centralised system cannot (Lebel et al., 2006). Polycentric institutional systems enable enhanced learning and experimentation and broader levels of participation and governance. This enables capitalisation on scale specific knowledge to enhance learning through information sharing, experience and knowledge across scales (Biggs et al., 2012).

Multi-layered institutions also provide for level dependent management interventions and mechanisms to address cross-level interactions without being detrimental to the capacity to self-organise at any level (Lebel et al., 2006). These arrangements contribute to the resilience of SESs through the provision of governance structures that facilitate other essential resilience enhancing principles, especially learning and adaptation as well as participation. It is not simply sufficient to establish a polycentric arrangement, however, social capital and maintaining or developing strong leadership are essential for successful polycentric governance (Biggs et al., 2012).

2.4   Achieving Sustainable Management of a Rural Social-Ecological Systems

The adoption of collective governance processes contributes to building rural resilience, which in turn provides sustainable governance and management of SESs, such as rural communities. By building resilient rural SESs, one can be hopeful for a sustainable future of the rural region. This is as the biophysical, societal and economic dimensions of the SES are all considered in cultivating a functioning community, which is equipped to adapt to changes in its external environment. This requires integrating the rural community with wider society, while safeguarding its heritage and unique characteristics as well as balancing traditions with new innovations. As such, how these resilience factors are incorporated into the revitalisation process is significant, as is how these factors are managed for a better balance and complementarity.

For example, too much social innovation risks the loss of traditional practices, however, if traditional practices are rigidly preserved, the village may not be able to function in modern society, creating an unsustainable situation. A balance has to be found between social innovation to modernise physical appearance and/or practices in the village, bringing in new settlers and developing new income streams, and safeguarding traditional practices and customs. Similarly, too much polycentricity in the governance and management of the SES, creates inefficiencies and confusion due to a lack of oversight. Whereas rigid and top down leadership risks the shortfalls of traditional command and control style approaches.

As such, the sustainability of a rural area requires recognition of the whole SES and the ability of that SES to cope with external disturbances. From the theoretical frameworks examined, it is evident that several common factors are influential in achieving rural resilience. In particular, the antecedent of social capital, processes of adaptive learning and leadership and a structure that involves bridging organisations and nested organisations or institutions are influential. How these resilience factors are balanced and implemented are explored in more detail in the following chapter.

References

Abel, N., Cumming, D. H. M., & Anderies, J. M. (2006). Collapse and reorganisation in social-ecological systems: Questions, some ideas and policy implications. *Ecology and Society, 11*(1), 17.

Adger, W. N. (2000). Social and ecological resilience: are they related? *Progress in Human Geography, 24*(3), 347–364.

Adger, N. (2003) Social capital, collective action, and adaptation to climate change. *Economic Geography, 79,* 387–404.

Adger, N. W. (2006) Vulnerability. *Global Environmental Change, 16*(3), 268–281. https://doi.org/10.1016/j.gloenvcha.2006.02.006.

Ambrosio-Albala, M., & Delgado, M. (2008). Understanding rural areas dynamics from a complex perspective. An application of prospective structural analysis. In *European Association of Agricultural Economists*. International Congress, August 26–29, Ghent, Belgium.

Ansell, C. & Gash, A. (2007) Collaborative Governance in Theory and Practice. *Journal of Public Administration Research and Theory, 18,* 543–571.

Avant, F., Rich-Rice, K., & Copeland, S. (2013). Leadership and rural communities. *International Journal of Business, Humanities and Technology, 3*(8), 53–59.

Baldwin, C., Smith, T., & Jacobson, C. (2017). Love of the land: Social-ecological connectivity of rural landholders. *Journal of Rural Studies, 51,* 37–52.

Banyai, C. (2009). Community leadership: Development and the evolution of leadership in Himeshima. *Rural Society, 19*(3), 241–261.

Berkes, F. (2002). Cross-scale Institutional Linkages: Perspectives from the bottom up. In: E. Ostrom, T. Dietz, N. Dolšak, P. C. Stern, S. Stonich, & E. U. Weber (Eds.), *The drama of the commons*. National Academy Press.

Berkes, F. (2009). Evolution of co-management: Role of knowledge generation, bridging organizations and social learning. *Journal of Environmental Management, 90,* 1692–1702.

Berkes, F. (2018). *Sacred ecology* (4th ed.). London/New York: Routledge.

Berkes, F., & Folke, C. (1992). A systems perspective on the interrelations between natural, human-made and cultural capital. *Ecological Economics, 5,* 1–8.

Berkes, F. & Folke, C. (1998) Linking social and ecological systems: Management practices and social mechanisms for building resilience. Cambridge: Cambridge University Press.

Berry, W. (1990). An argument for diversity. *Hudson Review, 42*(4), 537–548.

Biggs, R., Shluter, M., Biggs, D., Bohensky, E. L., Burnsilver, S., Cundill, G., Dakos, V., Daw, T. M., Evans, L. S., Kotschy, K., Leitch, A. M., Meek, C., Quinlan, A., Raudsepp-Hearne, C., Robards, M. D., Schoon, M., Shultz, L., & West, P. C. (2012) Towards principles for enhancing the resilience of ecosystem services. *Annual Review of Environment and Resources, 37,* 421–448.

Bingham, L. (2011). Collaborative governance. In M. Bevir (Ed.), *The SAGE handbook of governance* (pp. 386–401). SAGE.

Binns, T., & Nel, E. (2003). The village in a game park: Local response to the demise of coal mining in Kwazulu-Natal, South Africa. *Economic Geography, 79*(1), 41–66.

Bjorna, H., & Aarsaether, N. (2009). Combating depopulation in the northern periphery: Local leadership strategies in two Norwegian municipalities. *Local Government Studies, 35*(2), 213–233.

Bock, B. B. (2012). Social innovation and sustainability: how to disentangle the buzzword and its application in the field of agriculture and rural development. *Studies in Agricultural Eco-nomics, 114*(2), 57–63.

Bock, B. B. (2016). Rural Marginalisation and the role of social innovation; A turn towards Nexogenous development and rural reconnection. *Sociologia Ruralis, 56*(4), 552–573.

Bodin, Ö., Sandström, A., & Crona, B. (2017). Collaborative networks for effective ecosystem-based management: A set of working hypotheses. *Policy Studies Journal, 45*(2), 289–314.

Botes, L., & van Rensburg, D. (2000). Community participation in development: Nine plagues and twelve commandments. *Community Development Journal, 35*(1), 41–58.

Bourgon, J. (2010). The history and future of nation building? Building capacity for public results. *International Review of Administrative Sciences, 76*(2), 197–218.

Brondizio, E. S., Ostrom, E., & Young, O. R. (2009). Connectivity and the governance of multi-level social ecological systems: The role of social capital. *Annual Review of Environment and Resources, 34*(1), 253–278.

Brunckhorst, D. J., & Marshall, G. R. (2007). Designing robust common property regimes for collaboration towards rural sustainability. In A. Smajgl & S. Larson (Eds.), *Sustainable resource use: Institutional dynamics and economics* (pp. 179–207). Earthscan.

Bryson, J. M., Crosby, B. C., & Stone, M. M. (2015). Designing and implementing cross-sector collaborations: Needed and challenging. *Public Administration Review, 75*(5), 647–663.

Carpenter, S. R., Arrow, K. J., Barrett, S., Biggs, R., Brock, W. A., Crépin, A. S., Engström, G., Folke, C., Hughes, T., Kautsky, N., Li, C. Z., McCarney, G., Meng, K., Mäler, K. G., Polasky, S., Scheffer, M., Shogren, J., Sterner, T., Vincent, J., … de Zeeuw, A. (2012a). General resilience to cope with extreme events. *Sustainability, 4*, 3248–3259.

Carpenter, S. R., Folke, C., Norström, A., Olsson, O., Schultz, L., Agarwal, B., Blavenera, P., Campbell, B., Castilla, J. C., Cramer, W., DeFries, R., Eyzaguirre, P., Hughes, T. P., Polasky, S., Sanusi, Z., Scholes, R., & Spierenburg, M. (2012b). Program on ecosystem change and society: An international research strategy for integrated social–ecological systems. *Current Opinion in Environmental Sustainability, 4*(1), 134–138.

Carpenter, S. R., Westley, F., & Turner, M. G. (2005). Surrogates for resilience of social-ecological systems. *Ecosystems, 8*, 941–944.

Castro-Arce, K., Parra, C., & Vanclay, F. (2019). Social innovation, sustainability and the gover-nance of protected areas: Revealing theory as it plays out in practice in Costa Rica. *Journal of Environmental Planning and Management, 62*(13), 2255–2272.

Chambers, R. (1994). Participatory rural appraisal (PRA): Challenges, potentials and paradigm. *World Development, 22*(10), 1437–1454.

De Kraker, J. (2017). Social learning for resilience in social-ecological systems. *Current Opinion in Environmental Sustainability, 28*, 100–107.

De Silva, M. J., Huttly, S. R., Harpham, T., & Kenward, M. G. (2007). Social capital and mental health: a comparative analysis of four low income countries. *Social science & medicine, 64*(1), 5–20.

Dietz, T., Ostrom, E., & Stern, P. C. (2003). The struggle to govern the commons. *Science, 302*(5652), 1907–1912.

Emerson, K., & Nabatchi, T. (2015). *Collaborative governance regimes.* Georgetown University Press

Evans, L. S., Hicks, C. C., Cohen, P. J., Case, P., Prideaux, M., & Mills, D. J. (2015). Understanding leadership in the environmental sciences. *Ecology and Society, 20*(1).

Faulkner, L., Brown, K., & Quinn, T. (2018). Analyzing community resilience as an emergent property of dynamic social-ecological systems. *Ecology and Society, 23*(1), 24.

Folke, C., Carpenter, S., Elmqvist, T., Gunderson, L., Holling, C. S., & Walker, B. (2002). Resilience and sustainable development: Building adaptive capacity in a world of transformations. *Ambio, 31*, 437–440.

Folke, C., Carpenter, S., Walker, B., Scheffer, M., Chapin, T., & Rockstrom, J. (2010). Resilience Thinking: Integrating resilience, adaptability and transformability. *Ecology and Society, 15*(4), 20.

Folke, C., Hahn, T., Olsson, P., & Norberg, J. (2005). Adaptive governance of social-ecological systems. *Annual Review of Environment and Resources, 30*(1), 441–473.

Fulton, E., Smith, A., Smith, D., & van Putten, I. (2011). Human behaviour: The key source of uncertainty in fisheries management. *Fish and Fisheries, 12*, 2–17.

Fung, A., & Wright, E. (2001). Deepening democracy: Innovations in empowered participatory governance. *Politics & Society, 29*, 5–41.

Garmestani, A. S., & Benson, M. H. (2013). A framework for resilience-based governance of social-ecological systems. *Ecology and Society, 18*(1).

Gerlak, A. K., & Heikkila, T. (2011). Building a theory of learning in collaboratives: Evidence from the Everglades Restoration Program. *Journal of Public Administration Research and Theory, 21*(4), 619–644.

Glover, J. (2011). Rural resilience through continued learning and innovation. *Local Economy: The Journal of the Local Economy Policy Unit, 27*(4), 355–372. https://doi.org/10.1177/026909421 2437833.

Go, F. M., Trunfio, M. & Della Lucia, M. (2013) Social capital and governance for sustainable rural development. *Studies in Agricultural Economics, 115*(2), 104–110. https://doi.org/10.7896/ j.1220.

Gollagher, M., & Hartz-Karp, J. (2013). The role of deliberative collaborative governance in achieving sustainable cities. *Sustainability, 5*(6), 2343.

Gunderson, L., & Folke, C. (2005). Resilience—Now More Than Ever. *Ecology and Society, 10*(2), 22.

Hahn, T., Olsson, P., Folke, C., & Johansson, K. (2006). Trust-building, knowledge generation and organizational innovations: The Role of a bridging organization for adaptive comanagement of a wetland landscape around Kristianstad, Sweden. *Human Ecology, 34*(4), 573–592.

Harpham, T. (2008). The measurement of community social capital through surveys. In I. Kawachi, S. Subramanian & D. Kim, (Eds.), *Social Capital and Health*. New York, NY: Springer. https:// doi.org/10.1007/978-0-387-71311-3_3.

Heijman, W., Hagelaar, G., & Heide, M. (2007). Rural resilience as a new development concept, EAAE seminar Serbian Association of Agricultural Economists, Novi Sad, Serbia.

Herzberg, R. Q. (2020). Elinor Ostrom's governing the commons institutional diversity, self-governance, and tragedy diverted. *The Independent Review, 24*(4), 627–636.

Holling, C. S., & Meffe, G. K. (1996). Command and control and the pathology of natural resource management. *Conservation Biology, 10*, 328–337.

Hudson, R. (2010). Resilient regions in an uncertain world: wishful thinking or a practical reality? *Cambridge Journal of Regions, Economy and Society, 3*(1), 11–25.

Janssen, M. A., & Anderies, J. M. (2007). Robustness trade-offs in social-ecological systems. *International journal of the commons, 1*(1), 43–65.

Kallis, G., Kiparsky, M., & Norgaard, R. (2009). Collaborative governance and adaptive management: Lessons from California's CALFED Water Program. *Environmental Science & Policy, 12*(6), 631–643.

Lam, W. F. (2006). Foundations of a robust social-ecological system: Irrigation institutions in Taiwan. *Journal of Institutional Economics, 2*(2), 203–226.

Lam, W. F., & Chiu, C. Y. (2016). Institutional nesting and robustness of self-governance: The adaptation of irrigation systems in Taiwan. *International Journal of the Commons, 10*(2).

Leach, W. D., & Pelkey, N. W. (2001). Making watershed partnerships work: A review of the empirical literature. *Journal of Water Resources Planning and Management, 127*, 378–385.

Lebel, L., Anderies, J. M., Campbell, B., Folke, C., Hatfield-Dodds, S., Hughes, T. P., & Wilson, J. (2006). Governance and the capacity to manage resilience in regional social-ecological systems. *Ecology and society, 11*(1).

Leys, A. J., & Vanclay, J. K. (2011). Social learning: A knowledge and capacity building approach for adaptive co-management of contested landscapes. *Land Use Policy, 28*(3), 574–584.

Li, Y. R., Liu, Y. S., & Long, H. L. (2012). Characteristics and mechanism of village transformation development in typical regions of Huang-Huai-Hai Plain. *Acta Geographica Sinica, 67*(6), 771–782. (in Chinese).

Li, Y. R., Liu, Y. S., Long, H. L., & Wang, J. Y. (2013). Local responses to macro development policies and their effects on rural system in China's mountainous regions: The case of Shuanghe Village in Sichuan Province. *Journal of Mountain Science, 10*(4), 588–608.

Li, Y. R., Liu, Y. S., Long, H., Li, Y., Liu, Y., Long, H., & Cui, W. (2014). Community-based rural residential land consolidation and allocation can help to revitalize hollowed villages in traditional agricultural areas of China: Evidence from Dancheng County, Henan Province. *Land Use Policy, 39*, 188–198.

Li, Y., Westlund, H., & Liu, Y. (2019). Why some rural areas decline while some others not: An overview of rural evolution in the world. *Journal of Rural Studies, 68*, 135–143.

Lowe, P., Murdoch, J., & Ward, N. (1995). Networks in rural development: Beyond exogenous and endogenous models. In: J. D. van der Ploeg, & G. van Dijk (Eds.), *Beyond modernization: The impact of endogenous rural development*. Van Gorcum, Assen Binns & Nel, 2003.

Madsen, W., & O'Mullan, C. (2014). "Knowing me, knowing you": Exploring the effects of a rural leadership program on community resilience. *Rural Society, 23*, 151–160. https://doi.org/10.5172/rsj.2014.23.2.151.

Marsden, T. (2010). Mobilizing the regional eco-economy: Evolving webs of agri-food and rural development in the UK. *Cambridge Journal of Regions, Economy and Society, 3*(2), 225–244.

McGregor, D., Simon, D., & Thompson, D. (2006). The peri-urban interface in developing areas: The research agenda. In: D. McGregor, D. Simon, D. Thompson (Eds.), *The peri-urban interface: Approaches to sustainable natural and human resource use*. Earthscan.

Motesharrei, S., Rivas, J., & Kalnay, E. (2014). Human and nature dynamics (HANDY): Modelling inequality and use of resources in the collapse or sustainability of societies. *Ecological Economics, 101*, 90–102.

Moulaert, F., MacCallum, D., Mehmood, A., & Hamdouch, A. (2013). General introduction: The return of social innovation as a scientific concept and a social practice. In F. Moulaert, D. MacCacllum, A. Mehmood, & A. Hamdouch (Eds.), *The international handbook on social innovation: Collective action, social learning and transdisciplinary research* (pp. 1–9). Edward Elgar.

Murdoch, J. (2000). Networks: A new paradigm of rural development? *Journal of Rural Studies, 16*(4), 407–419.

Narain, V., & Vij, S. (2016). Where have all the commons gone? *Geoforum, 68*, 21–24.

Nardone, G., Sisto, R., Lopolito, A. (2010). Social capital in the LEADER Initiative: a methodological approach. *Journal of Rural Studies, 26*(1), 63–72. https://doi.org/10.1016/j.jrurstud.2009.09.001.

Normann, R. (2013). Regional leadership: A systemic view. *Systemic Practice and Action Research, 26*, 23–38.

Olfert, M. R., & Partridge, M. D. (2010). Best practices in twenty-first-century rural development and policy. *Growth and Change, 41*(2), 147–164.

Olsson, P., Folke, C., Galaz, V., Hahn, T., & Schultz, L. (2007). Enhancing the fit through adaptive co-management: creating and maintaining bridging functions for matching scales in the Kristianstads Vatternrike Biosphere Reserve, Sweden. *Ecology and Society, 12*(1), 28.

Olsson, P., Galaz, V., & Boonstra, W. J. (2014). Sustainability transformations: A resilience perspective. *Ecology and Society, 19*(4), 1.

Olsson, P., Gunderson, L. H., Carpenter, S. R., Ryan, P., Lebel, L., Folke, C., & Holling, C. S. (2006). Shooting the rapids: Navigating transitions to adaptive governance of social-ecological systems. *Ecology and Society, 11*(1), 18.

Pahl-Wostl, C., Craps, M., Dewulf, A., Mostert, E., Tabara, D., & Taillieu, T. (2007). Social learning and water resources management. *Ecology and society, 12*(2).

Parra, C. (2013). Social sustainability, a competitive concept for social innovation? In F. Moulaert, D. MacCallum, A. Mehmood, & A. Hamdouch (Eds.), *The international handbook on social innovation: Collective action, social learning and transdisciplinary research* (pp. 142–154). Edward Elgar.

Provan, K. G., & Kenis, P. (2008). Modes of network governance: Structure, management, and effectiveness. *Journal of Public Administration Research and Theory, 18*(2), 229–252.

Rammel, C., Stagl, S., & Wilfing, H. (2007). Managing complex adaptive systems—A co-evolutionary perspective on natural resources management. *Ecological Economics, 63*(1), 9–21.

Ratner, B. D., & Allison, E. (2012). Wealth, rights and resilience: An agenda for governance reform in small-scale fisheries. *Development Policy Review, 30*(4), 371–398.

Redman, C. L. (2014). Should sustainability and resilience be combined or remain distinct pursuits? *Ecology and Society, 19*(2), 37.

Ryser, L., & Halseth, G. (2010). Rural economic development: A review of the literature from industrialized economies. *Geography Compass, 4*(6), 510–531.

Salvia, R., & Quaranta, G. (2017). Place-based rural development and resilience: A lesson from a small community. *Sustainability, 9*(6), 889.

Schouten, M., van der Heide, M., & Heijam, W. (2009). *Resilience of social-ecological systems in European rural areas: Theory and prospects.* Paper prepared for presentation at the 113th EAAE Seminar: The role of knowledge, innovation and human capital in multifunctional agriculture and territorial rural development, Belgrade, Republic of Serbia December 9–11, 2009.

Scott, M. (2013). Resilience: a conceptual lens for rural studies? *Geography Compass, 7*(9), 597–610.

Singh, A. K., & Narain, V. (2020). Lost in transition: Perspectives, processes and transformations in Periurbanizing India. *Cities, 97,* 102494.

Stewart, J., Tyler, M. E. (2019). Bridging organizations and strategic bridging functions in environmental governance and management. *International Journal of Water Resources Development, 35*(1), 71–94. https://doi.org/10.1080/07900627.2017.1389697.

Tenza, A., Perez, I., Martinez-Fernandez, & Gimenez, A. (2017). Understanding the decline and resilience loss of a long-lived social-ecological system: Insights from system dynamics. *Ecology and Society, 22*(2), 15.

Titumir, R. A. M., Afrin, T., & Islam, M. S. (2020). Traditional knowledge, institutions and human sociality in sustainable use and conservation of biodiversity of the Sundarbans of Bangladesh. In *Managing Socio-ecological production landscapes and seascapes for sustainable communities in Asia* (pp. 67–92). Singapore: Springer.

van der Ploeg, J. D., & Marsden, T.K. (2008). *Unfolding webs: The dynamics of regional rural development.* Van Gorcum.

Vij, S., Narain, V., Karpouzoglou, T., & Mishra, P. (2018). From the core to the periphery: Conflicts and cooperation over land and water in periurban Gurgaon, India. *Land Use Policy, 76,* 382–390.

Walker, B., Gunderson, L., Kinzig, A., Folke, C., Carpenter, S., & Shultz, L. (2006). A handful of heuristics and some propositions for understanding resilience in social ecological systems. *Ecology and Society, 11*(1), 13.

Walker, B., & Salt, D. (2006). *Resilience thinking: Sustaining ecosystems and people in a changing world.* Island Press.

Walker, B., & Salt, D. (2012). *Resilience practice.* Island Press.

Walters, C. J. (1986). *Adaptive management of renewable resources.* Macmillan Publishers Ltd.

Walters, C. J., & Holling, C. S. (1990). Large-scale management experiments and learning by doing. *Ecology, 71*(6), 2060–2068.

Wamsler, C. (2017) Stakeholder involvement in strategic adaptation planning: Transdisciplinarity and co-production at stake? *Environmental Science & Policy* 75 148–157 https://doi.org/10.1016/j.envsci.2017.03.016.

Wang, M. (2020). The role of local knowledge for rural revitalization in China: Social-ecological lessons learned through disasters, architecture, and education. *Bioregional Planning and Design: Volume II*, 259–278.

Warner, M. (1999). Social capital construction and the role of the Local State. *Rural Sociology, 64*(3), 373–393. https://doi.org/10.1111/j.1549-0831.1999.tb00358.x.

Westley, F. R., Tjornbo, O., Schultz, L., Olsson, P., Folke, C., Crona, B., & Bodin, Ö. (2013). A theory of transformative agency in linked social-ecological systems. *Ecology and Society, 18*(3), 27.

Zhang, Y., He, D., Lu, Y., Feng, Y., Reznick, Y. (2013). The influence of large dams building on resettlement in the Upper Mekong River. *Journal of Geographical Sciences, 23*(5), 947–957. https://doi.org/10.1007/s11442-013-1054-2.

Chapter 3
Building Rural Resilience

Abstract Rural resilience is found to be supported by the presence of social capital, effective learning and adaptation, social innovation, leadership during the collaboration process, a dynamic bridging organisation and polycentric institutional arrangements. A deeper understanding of how these factors relate to rural resilience is developed through an in-depth analysis of the LCW case study. The role of the Programme team as a bridging organisation is particularly pertinent in this case due to the initial disengagement of the wider Hong Kong society in rural areas and the rural population being too depleted to drive revitalisation themselves. This required a new governance model, which took advantage of the growth in civil society, driven by the Programme team to establish new collaborative approaches to rural revitalisation.

3.1 How to Build Rural Resilience

The previous chapter identified that rural resilience is supported by the presence of social capital, effective learning and adaptation, social innovation, leadership during the collaboration process, a dynamic bridging organisation and polycentric institutional arrangements. This chapter aims to develop a deeper understanding of how these factors relate to rural resilience through an in-depth analysis of the LCW case study. All of the factors identified in Chapter two are featured in the revitalisation process and have played important roles in building rural resilience of the LCW area. Each chapter will feature a set of problems and obstacles pertaining to the revitalisation of LCW followed by explanations of the ways in which the factors have been integrated into the revitalisation approach and how they contribute to addressing the problems as well as the building of rural resilience.

Each of the following three Sects. (3.2–3.4) will focus on one or two of the key factors contributing to rural resilience:

3.2   Learning and adaptation.
3.3   Social capital and social innovation.
3.4   Polycentric institutional arrangements.

The role of PSL as a bridging organisation drove the revitalisation process and sought to ensure that the process built rural resilience in the LCW area. Within this

© University of Hong Kong 2021   39
J. M. Williams et al., *Revitalising Rural Communities*, SpringerBriefs on Case Studies
of Sustainable Development, https://doi.org/10.1007/978-981-16-5824-2_3

role, PSL instigated adaption and learning, built social capital, cultivated social innovation and implemented polycentric institutions and participatory governance structures. PSL also provided leadership, especially in the building of new socio-economic models and experimentation of initiatives such as sustainable farming practices. The goal was to discover and establish feasible models with the community that could eventually be taken up by the community. Focus was on building a community, bringing together interests from Indigenous villagers and new settlers when developing models for the rural revitalisation of LCW, and so PSL was constantly having to bridge these interests. PSL also had to liaise between the villagers and the government and green groups in its role as a bridging organisation, creating venues and organising meetings that neither group alone would have been able to instigate. As such, the role of PSL as a bridging organisation and the provision of leadership is examined within each of the subsections.

The need for a bridging organisation in the LCW case was due to the Hong Kong context. While the general trend for revitalisation approaches elsewhere is to seek to empower local communities and have them as the major impetus and driving force for revitalisation, this is not readily available in Hong Kong. Outmigration in the 1960s and 1970s means that many rural areas do not possess a sufficient population or the necessary social capital to be able to drive revitalisation efforts. More significantly, however, is the lack of diversified connection in Hong Kong between urban and rural communities. Despite the interconnections between urban and rural areas in Hong Kong, urban communities tend to consider rural affairs as being private issues, concerning only the Indigenous population. Rural areas are often equated with country parks and so are recreational spaces to be protected rather than part of Hong Kong's cultural landscape. Many also tended to hold a negative perception towards Indigenous villagers due to their right to land, a scarce commodity in Hong Kong.

As such, rural development was not a prominent topic in public debates and so the wider community believed that rural areas were not related to them, even the younger generations of Indigenous villagers felt no, or only a limited, connection with rural development. The more recent interest in localism and self-sufficiency has, however, begun to pique some interest in local food production as well as the mental benefits of outdoor and green spaces. As such, the urban–rural connections, which have been weakened, are starting to be recognised.

Consequently, a new governance model was required to shift rural areas from being seen as private issues to one that concerned the public. The empowerment of civil society in Hong Kong provided a new voice in the matter and a means to reconnect the public with rural areas. As such, the involvement of civil society actors, in this instant PSL in particular, as bridging organisations between the urban and rural communities was essential. These actors provided a departure from the old governance models and worked to establish new collaborative approaches to rural development.

Rural and urban areas are economically, socially and environmentally interlinked, with each space benefiting from these interlinkages. Without rural development, urban development generally cannot occur, especially where agriculture is the mainstay of the economy. Urban areas import nearly all their ecosystem services from rural areas, relying on them to meet their demand for food, water, wood, raw materials and other resources. Making rural areas a necessity for the functioning of urban areas. In return, rural areas gain markets, farm inputs and employment opportunities from urban areas. As such, the interface between rural and urban areas should be managed to ensure that urban development does not adversely impact rural ecosystems and rural life. Rather, rural populations and ecosystems should be supported and protected for their sustainable service delivery (Gebre & Gebremedhin, 2019).

While the LCW situation is unusual, it is not necessarily unique. Peri-urban areas elsewhere in the world have suffered similar deterioration of urban/rural ties due to outmigration and globalisation, weakening local markets and economic opportunities in rural areas. For example, a similar context may be present in parts of rural China, where the working population has migrated in search for better paying jobs, leaving rural areas populated by the elderly and young. Similarly, in Europe, the youth have left rural villages seeking better opportunities due to the inaccessibility of their villages. As such, this model could be applicable, or at least provide insights, in addressing rural issues elsewhere in the world.

3.2 Learning and Adaptation

3.2.1 Introduction: Balancing the Competing Needs and Interests of the SES

Policy issues, particularly those regarding the environment and natural resource management, are predisposed to significant uncertainties. Knowledge regarding the nature, magnitude or likelihood of adverse consequences from certain activities is often undetermined or incomplete. Rural revitalisation is no exception as it involves human interaction with the natural environment, ecosystem services and the management of natural resources. In the hopes of building sustainable and resilient rural societies, it is essential to develop and conserve the integrity of SESs. This includes maintaining biodiversity and resource efficiencies, while minimising negative externalities of human activities. Adaptation and learning are, therefore, necessary to protecting SES integrity in the face of internal and external shocks and stresses. This is particularly pertinent to developing social-ecological integrity for rural communities in the peri-urban context given the dynamism caused by urban influences. Maintaining the ability to learn and adapt is then not only needed during the revitalisation process but also for the resilience of rural communities in the longer term.

It is increasingly evident that human activities, such as urbanisation, land use change and resource exploitation, are threatening global diversity, with up to a million species at risk of extinction (Grooten & Almond, 2018; IPBES, 2019). Sustainable stewardship of the rural environment, and particularly the peri-urban areas that are under greater threats and pressure, is, therefore, crucial to environmental and societal well-being. This is, however, extremely challenging as climate change causes the frequency and intensity of environmental hazards such as typhoons, rainstorms and droughts to increase, adding to uncertainty in resource management. To increase communities' capacities to respond, or even thrive in the face of uncertainties, efforts are made from within or from external organisations to build communities' learning and adaptation capacities. PSL in this case acted as a bridging institution bringing stakeholders together and providing leadership in encouraging cross-sector learning and building LCW's adaptive capacity.

Restoring SES integrity in a largely abandoned rural community, as was the case for LCW, is no easy task, as abandonment resulted in the disruption in the human/nature balance of the area. While in its hey-day, LCW may have struggled to maintain its population size without detrimentally impacting the environment, the current loss of population and human intervention has also had a negative effect on the SES. Social heritage and culture were in danger of being lost while habitats and ecosystems that had been created and protected by rural life were degrading. Given the dilapidated state of the village, new actors, organisations or individuals, must be drawn upon to contribute towards the goal of revitalisation. To achieve this, the Programme needs to take advantage of the peri-urban context, improving the feasibility for individuals to make regular day trips to work in the village combined with a range of initiatives to attract people with different skills and interests in rural culture and life. At the same time, all stakeholders were cautious of protecting the village from being overwhelmed by urban influences or flow of people.

In terms of resource management, the village also suffered due to its relative remoteness, relying on old septic tanks and waste was often disposed of improperly, with negative environmental impacts. Revitalisation efforts, therefore, needed to focus on a combination of rediscovering sustainable traditional practices and reinventing sustainable management and resource use approaches appropriate to current conditions and contextual factors. Exploring the case of water resource management in this chapter (further discussions on the related issues and conflicts in this area can be found in Sect. 3.4.5) demonstrates that the resumption of traditional management approaches amongst Indigenous villagers was facilitated through a new local platform that was established amongst organisations working in LCW, new settlers/farmers and Indigenous villagers. Such efforts seek to find the equilibrium between the social and ecological sides of the system. Interventions as a part of the Programme look to optimise all stakeholders' knowledge on human-nature interactions, adjusting their actions and roles to maintain the ecological integrity in the local and territorial level while cultivating a thriving social community. This process entails significant adaptive and flexible learning that was, in this case, facilitated by the bridging organisation.

3.2.2 Redeveloping Socio-ecological System Integrity Through Collaborative Learning

The SES framework emphasises the interdependency between social and bio-physical systems as well as between SESs at different scales. LCW is a complex SES, both its existing and historical bio-physical and social sub-systems had to be understood and respected throughout the revitalisation process. Learning about the current and historical SESs were necessary before constructing a new SES to maintain the integrity of the community.

In terms of the existing bio-physical sub-systems, LCW and the surrounding area are recognised as a site of ecological importance in Hong Kong. The area houses wetlands and mangroves, a seagrass population of scientific interests, a marine park and a Fung Shui forest with Special Area status due to its diversity of plant species. In Hong Kong, the common approach to conservation is to draw a clear divide between human activities and nature. While this is popular with some conservationists, the ecological value of farming village landscape, which is a product of long-term inter-action between nature and humans, is often neglected. Consequently, the revitalisa-tion programme placed an emphasis on the idea that cultural and natural values are interrelated and shape each other over time.

In rural communities, traditional practices and lifestyles tend to interact closely with biodiversity. Approaches to resource management are often refined by local farmers and villagers over decades. A decline in biosphere integrity and ecosystem degradation pose particular threats to rural regions, weakening their resilience and risk compromising the ability of these communities to respond to natural disasters and mitigate the impacts of climate change. Healthy, diverse and productive ecosys-tems deliver important ecosystem services that rural communities rely on for both on-farm (e.g. agriculture) and off-farm (e.g. tourism) economic activities (Roe et al., 2019). Reconnecting ecological and social systems can maintain sustainable liveli-hoods, alleviate poverty and foster long-term sustainability. Conserving biodiversity and maintaining rural livelihoods are interdependent and global goals of achieving sustainable development could be significantly hampered by the current trajectory of rural biodiversity loss.

During the revitalisation process, balance between human activities and ecolog-ical conservation had to be struck to ensure the SES's integrity. Especially at the beginning of the Programme, there was concern from citizens and green groups about the removal of vegetation for the restoration of farmlands and the potential impact of farming activity on the local biodiversity. Concerns about the possibility of increased urban influences leading to land and resources of villages being exploited for private development are common in the peri-urban context. Given Hong Kong's socio-economic circumstances, with extremely high land value and numerous cases of village land in the urban fringe being sold to private developers, public scepticism towards the Programme was understandable. The Programme team leveraged its role as a mediator to ensure these concerns were properly addressed. They did this by ensuring a transparent process, strengthening communication with concerned groups and relevant government departments, particularly in the initial stages of farming revitalisation. Results of baseline ecological surveys and an agricultural revitalisation plan were shared with concern groups. Through several liaison meetings, comments were collected, and concern groups were able to participate in the Programme team's learning journey for devising and implementing the revitalisation plan.

Revitalisation efforts rejuvenated LCW's farmland alongside ecologists with local experience in agro-ecosystem management to raise productivity while conserving biodiversity. The creation of a properly managed farming habitat has resulted in an increase in overall biodiversity by enhancing the sites' habitat heterogeneity. As a result, the Chinese Bullfrog, a class II protected species in China[1] that used to be commonly found in the farmland, has returned since farming was revitalised. Carbon emissions due to vegetation removal were minimised through the production of biochar and using solar powered fences as physical pest control. The team has adopted sustainable agricultural practices, conducted experiments and studies on eco-agriculture methods, to minimise negative impacts and to enhance the quality of habitats.

Besides engaging with concern groups and working with ecologists, the Programme team utilised opportunities made available by improved accessibility to urban areas and conducted a range of initiatives to generate interest amongst the wider society. These initiatives led to collaboration between local community members and interested individuals who are willing to support the LCW community to build and sustain the various SESs. Villagers and new settlers were engaged in planting incense trees and resuming paddy farming as well as protecting the landscape such as the Feng Shui forest and the woods surrounding the village. Concepts of maintaining socio-ecological integrity were advocated by the Programme team through daily interactions and numerous informal/semi-formal meetings with villagers and new settlers. Most of the villagers are open to new knowledge and eager to learn about sustainable practices in farming and in their daily lives.

Given that agricultural revitalisation was a key component of the Programme and of interest to the organisations involved in the revitalisation process, attention was paid to maintaining the ecosystem services provided by the rich natural environment of the area while increasing food productivity. Through the careful design of all its initiatives and engagement with partnering organisations and community members, the Programme maintained the aim of building a sustainable rural community, rather than revitalisation of the economy or agriculture. Increased economic vitality and agricultural production were part of the means to that aim. This mind-set rules out large-scale agricultural development driven by principles of economic efficiency and/or technological advancement, which is often a risk in peri-urban areas, as demonstrated in India (Narain & Vij, 2016). Sustainable agriculture and land management were adopted to promote greater species diversity, increase overall

---

[1] Chinese bullfrog (虎纹蛙), *Rana tigrine* is included as one of China's national key protected wild animals by the State Forestry and Grassland Administration: http://www.forestry.gov.cn/main/3954/20180104/1063883.html.

productivity and ensure climate resilience (King, 2018). The resource and productive efficiency of small-scale local food production has been demonstrated in numerous developing countries and rural regions. Agricultural diversification and employing agroforestry approaches is key to increase resilience to climate change and environmental shocks. Therefore, small-scale food production and agroforestry approaches were employed in the Programme. Following the principles of agroforestry, the team was able to reduce tree felling and create alternative habitats, which allows a balance between agricultural activities and the protection of the natural environment. Other farming practices adopted include organic farming as the principal production mode on the revitalised farmlands, which prevents chemical pollution while enhancing biodiversity and soil fertility.

Eco-agriculture model

Several actions were taken to create a more resilient and productive system:

1. A new farming community that embraced the principles of organic farming and eco-friendly agriculture was created. PSL recruited and taught volunteers organic farming methods and instigated the Hong Kong Organic Resource Centre's[2] organic standards as well as other rules designed to protect the wildlife and local habitats.

2. A small-holder farming approach that encouraged the establishment of "tiny-scale" farms at the local level was adopted. Different farms choose to grow different crops and so agricultural biodiversity is improved, which helps prevent large-scale pest problems. Information exchange is also encouraged within the farming community to facilitate building community resilience.

3. Implementing agroforestry to diversify the production modes in LCW as well as Hong Kong. The Programme developed an agroforest coffee and native forest species model to serve production and conservation purposes. The native forest trees provide shade and shelter for the coffee trees as well as moderate temperature.

Besides engaging in adaptive and collaborative learning in relation to farming activities, these processes also took place in other aspects of reconstructing a sustainable community in LCW. Taking advantage of the peri-urban context, attempts were made to tap talents from the urban areas for LCW. Funding schemes were designed as part of the Programme to attract specialist groups to further develop diverse opportunities and socio-economic models, which become a part of the puzzle to uphold the socio-ecological integrity of the community. Through these initiatives, specialist groups such as artists and Indigenous villagers engage in the rediscovery of traditional cultural or farming practices. This led to innovative and sustainable ways of

---

[2] Hong Kong Organic Resource Centre: http://hkorc-cert.org/en/.

using local resources, such as native plant species and the development of personal care products, soil-based paint and natural dye.

3.2.3   Effective Monitoring and Adaptive Planning

To ensure the integrity of the SES was upheld during the LCW revitalisation process, it was first important to understand the current SES. This was done by conducting a baseline ecological survey of the natural resources (water, biodiversity, land) at the village and landscape level at the start of the ecological revitalisation process. The aims of the baseline ecological survey were to provide background information of the target area and assist decision-makers in formulating priorities for development.

The current biodiversity and ecological patterns were analysed to reveal the dominant flora and fauna, local biodiversity hotspots and species of conservation importance. Important species and habitats included trees with a DBH[3] over 150 mm and of conservation value, are protected under the project, and water ferns, which are classed as vulnerable (category II),[4] were relocated. Any revitalisation activity had to be conducted relative to and measured against the baseline conditions.

Biodiversity monitoring remains an on-going process, it informs the project team, the local community and the wider community of interest about the biodiversity status of the site and helps determine the success of interventions. It also allows for adaptation and learning to occur as the project team can quickly identify if revitalisation actions are having a negative impact on the ecosystem and make the necessary adjustments.

Ongoing biodiversity monitoring

Farming practices and management strategies were adjusted by the Programme to attain both livelihood and conservation objectives:

- Throughout the revitalisation process, the community is encouraged to report any sightings of rare or important wildlife as well as wildlife accidents in the village through Whatsapp. This enables the Programme team and the community to act responsively and collectively to protect wildlife and adjust interventions.
- Biodiversity monitoring is also integrated into agroforestry development to evaluate the conservation performance of different agroforestry treatments (e.g. frequency of weeding, use of mulch and crop cover) so that these can be modified according to the data.

---

[3] Diameter at breast height.

[4] Water fern, *Ceratopteris thalictroides,* is classed as vulnerable according to China's national key protected wild plants issued by the State Forestry and Grassland Administration: http://www.for estry.gov.cn/main/3954/20180925/143410933280757.html.

- Key habitats for freshwater invertebrates and rice fish were identified and by working with the farmers, new pools and a freshwater marsh were designed within the project area to protect the concerned species.
- The farmers explained the whole farming operation processes to the scientists involved in the project to identify key actions, factors and causes of changes to local biodiversity due to farming activities.

As is often the case in peri-urban areas, changes in population and livelihoods lead to the emergence of issues regarding common resource management. This was no exception in LCW, especially in the case of water resource management. Traditionally, stream sediments were regularly cleared by the Indigenous villagers to maintain the appropriate stream depth and irrigation structures. As villagers migrated from the village, the practice ceased, resulting in sediment accumulation and a shallow stream. Over time, the topological changes lowered the stream's capacity to regulate flooding and seasonal drought. Stream banks tend to become seriously damaged during intense rainstorm events.

To investigate the relationship between rainfall and streamflow as well as evaluate the water resources in the catchment, the Programme engaged local hydrology experts and academics to conduct hydrology simulation and modelling.

Whole catchment management approach

Previously, the hydrological system at LCW was a human-nature interactive system. Villagers built a reservoir upstream and masonry banks to protect downstream irrigation channels and farmlands. The result was an extensive irrigation system that watered thousands of terraced farmlands.

Water resources, therefore, cannot be managed in one locality, a whole catchment approach is required as the village uses natural water sources for domestic and agricultural use.

The hydrology system had to be re-learnt to account for the more environmental orientation of the modern system. Research provided information about both the environmental and human factors of the current hydrological system. It was found that farming revitalisation has not induced significant or persistent negative impacts on the system but the data does suggest the need for flood and drought management.

Research helped the Programme team to formulate 'evidence-based suggestions' for the relevant government departments. Misunderstandings of traditional practices in society, such as regular de-silting having a negative ecological impact, were also remedied.

With the resumption of farming activities, there is a need to maintain the river banks, reservoir and irrigation channels. A new platform was formed to manage farming related affairs between all the farming groups, which includes Indigenous villagers and new villagers. During regular meetings, they make decisions on issues such as the frequency and date of collective maintenance work for water ways and the scope and schedule of regular duties for each farming group. The farming community removes sediment on an annual basis using manual effort and small machines to transport sediment accumulated around the dam to the lower stream. This temporarily raises the water storage capacity for irrigation during more extreme droughts but does not aid flood prevention. To allow faster discharge of flood water, dredging and widening of drainage channels within farming areas was conducted, which have helped alleviate the severity of floods.

New institution, such as the Community farmers' meeting, facilitates collaborative management of commons in the village while at the same time, helps to enhance community resilience. The capacity of local community members to take precautionary measures and implement recovery actions, especially in relation to managing the farms at LCW, has improved since the revitalisation project has been implemented. The Programme team and staff at the Conservancy Association who have also been working in LCW observed that community members have developed a tendency to interact more intensively before and after typhoons. They inform each other of who is residing in the village and provide updates on the condition of the village or farms. Some community members have attributed these changes to the Programme team and other organisations working in LCW. The village heads have also contributed by providing a form of leadership that reminds community members of issues that need attention, for example clearing the drainage channels adjacent to farming areas and preparatory actions that may minimise the impact of typhoons. Eventually, villagers developed the habit of reminding each other and working together to implement these precautionary or recovery actions. During periods of drought, some villagers gather to dredge parts of the drainage channels and they coordinate as a community to try to distribute the irrigation water more evenly between the farms.

Adaptive planning also had to occur when revitalising farming at LCW. Initially, the Programme facilitated the reinstatement of rice paddy farming. It soon became clear, however, that while traditionally a major agricultural output of the village, rice farming was now problematic. Ventilation around the village is poor and humidity levels are high due to the dense surrounding vegetation, which provides favourable conditions for rice disease. The vegetation also attracts birds, which inflict serious damage on rice crops. The heavy metal content of the soil has also accumulated since the abandonment of farming and so is naturally high, consequently, the harvested rice contains trace amounts of arsenic, which was undesirable.

The Programme learnt from this experience and adapted its approach to farming. Trials were conducted to select more suitable rice varieties that are disease resistant, have a low absorption of heavy metals while still yielding good quality grains. Rice farming was also shifted from the main agricultural produce to being maintained for conservation and demonstration purposes. Instead, more suitable crops are now being farmed.

Adaptive farming

Crop choices were adapted to suit the environment:

Flood tolerant crops such as lotus are grown in areas that are prone to frequent flooding.

As the traditional irrigation system has not been fully restored due to insufficient manpower, less water consuming crops such as coffee, ginger, turmeric and turnips are grown to lower water demand.

Crops that are vulnerable to being eaten by wild boar, such as peanuts and sweet potatoes, are avoided to lower the chance of human-wildlife conflicts.

The Programme team is also experimenting with the use and propagation of native species (e.g. incense trees and Hakka sweet tea) to innovate new crop products that are more adaptive to the native ecosystem.

3.2.4   Lessons and Conclusions

Revitalisation efforts have struck a balance between agricultural activities and the protection of the natural environment to preserve the SES's integrity and construct a sustainable community, which has built a resilient rural system. This was done by utilising adaptation and learning process as well as leveraging the Programme team's role as a bridging institution to bring in expert and local knowledge, ease concerns and ensure ecosystem protection. This contributed to building the rural resilience of the area and so for the Programme to sustainably develop the village.

Continuous learning and adaptation have allowed the Programme to be flexible in incorporating the social and environmental needs of the SES while working with the community to restore the functionality of the village. The baseline survey and continuous biodiversity monitoring plays a key role in ensuring the integrity of the ecological system is maintained. Local knowledge identified how flora and fauna was used traditionally in the village, providing essential information for biodiversity exercises. PSL's role in bringing scientists and Indigenous villagers together facilitated the sharing of knowledge to build a more complete picture of the area's historical, current and evolving ecological system. As a result, the Programme team and villagers are sensitive to changes in the environment and can respond quickly and appropriately.

The employment of agroforestry also constitutes an important component of the sustainable revitalisation approach. The agroforestry model integrates the production of crops with the conservation of habitats, which were previously considered to be mutually exclusive. Thus, the implementation of this model at LCW offers an example of an innovative approach to achieve both purposes simultaneously. By creating a diversity of habitats and crops, it also helps to build socio-ecological resilience. This adaptive approach to restoring agriculture also allowed for the local need for economically productive agricultural activities to be met by the new environment. This was done through learning from local and scientific knowledge and experimenting with different crops to find those best suited to the changed environment. As a result, rice production was scaled back and growing crops such as coffee, ginger and turmeric were found to be more economically and environmentally sustainable.

Adapting to the new hydrological system of the area was also important in ensuring a resilient rural system. The Programme brought in scientists and hydrological experts, liaised with government officials and worked with the villagers to develop a suitable management approach for the region's water resources. The villagers' knowledge of the stream, the irrigation system and their traditional management was vital in understanding the historical hydrological system and in addressing flooding and drought issues. Disuse of the irrigation system and the designation of the stream as ecologically important in 2015, however, meant that traditional approaches had to be adjusted. PSL was able to take the lead in finding solutions for managing the river by bringing in scientists and liaising with the involved stakeholders.

While approaches adopted in agricultural revitalisation at LCW has been characterised by intensive learning and adaptation, it is important to note that maintaining stability in other aspects such as the natural and cultural landscape of the village was also integral to rural sustainability. The Feng Shui forest and the mangroves continue to be protected. This landscape, with its transition from landward subtropical forest to seaward seagrasses, found at LCW has been identified as likely the most intact mangrove in China (Morton, 2016). Other than its environmental significance, the conservation of traditional agricultural landscapes, especially those in Asia, have been increasingly vulnerable under rapid urbanisation and possess great cultural value. The natural landscape (previously maintained by the Indigenous villagers for reasons including Feng Shui, sustaining provision of natural resources, providing stable climatic condition for settlement, and protection against waves and landslides) integrated with the setting of the traditional Hakka village in LCW warrants it an outstanding example of an organically evolved cultural landscape in Southern China and in Hong Kong (Lam et al. 2018).

As demonstrated in this subsection, building a sustainable SES with its inherently complex integration of bio-physical and social sub-systems calls for an adaptive and flexible approach. This allows adjustments to be made at ease and as frequently as required. Such adjustments occur as learning takes place through a combination of recovering Indigenous knowledge, identifying relevant scientific information and on-site experimentation. Altogether, these factors inform decisions on changes to be made to management approaches and practices at the local community level, and lead to developing new knowledge as reference for other rural communities.

3.3 Social Capital and Social Innovation

3.3.1 Introduction: The Purpose of Social Innovation for Rural Revitalisation

Rural areas in decline require innovative diversification strategies that can take advantage of unique local resources to introduce new socially and economically viable activities. A vibrant rural economy weakens 'push factors' that drives out-migration, encourages rural innovation and empowers the rural youth (Vargas-Lundius& Sutti, 2014). Villages in the peri-urban context have even greater potential to become valuable breeding grounds for experimentation of alternative ways of living for the urban population. Aligned with skills training and capacity building in a variety of areas, economic diversification accelerates inclusive growth and drives rural productivity. The availability of diverse skill sets and growth drawn from the urban areas contributes to the communities' abilities to deal with problems, hence its resilience.

Revitalisation, in a holistic manner, builds economic vibrancy for the broader goal of incubating social capital and cultivating social innovation to enhance rural resilience. This is as economic vibrancy does not exist on its own and must be built upon a dynamic society, where by enhancing the health and well-being of the community, their access to resources and services to build sustainable livelihoods is ensured. The building of both economic vibrancy and resilience for a village often involves enhancing the interlinkages between rural and urban areas. Such interlinkages, in many instances, represent the ways in which the two have developed a mutual dependency on each other.

Given its state of near abandonment, the building of rural resilience at LCW was deemed immensely challenging. The out-migration of almost the entire Indigenous population due to the lack of livelihood opportunities at the village led to a loss of social and cultural capitals. It was clear that although some Indigenous villagers were willing to return to the village and be involved in the revitalisation process, not enough were present for the rejuvenation of the village. There was an urgent need for a community network to support physical restoration works and to establish the village community and livelihoods. A constellation of initiatives were built into the LCW revitalisation programme to stimulate social innovation with the intention that such innovations will bring improved connections and collective empowerment. Enhancing the local community's connections internally and externally, in terms of socio-political levels and spatial scales, and its capacity to organise collective action supports the establishment of a proactive and sustainable SES. One of the key principles running through all of these initiatives was the involvement of community members. Any creation of new farm-related or non-farm ventures and enterprises are developed in close cooperation with, and reflect the needs and aspirations of, local stakeholders to strengthen their agency and social capital.

Social capital assets were increased through the creation of employment as well as economic and learning/training opportunities. This was done through placemaking of the rural village, whereby rural based socio-economic models were developed as social innovation strategies to engage both villagers and urban dwellers in a collaborative process. This built new social capital to support rural–urban symbiosis and enhanced the sustainability of the wider city. The village community also had to be supported and safeguarded to ensure the bonds between the villagers and between the villagers and the environment are respected and maintained to preserve the structural social capital of the village.

3.3.2 Institutional and Resource Challenges to Social Vibrancy

Despite being aware of the need to draw upon a wider network of people, organisations and resources to support the revitalisation efforts, and the value of building rural–urban linkages, there were several key challenges that needed to be overcome.

Firstly, the divide between rural and urban communities, coupled with the complexities surrounding rural land use and outmigration in Hong Kong in the 1960s–70s, meant that the rural, as well as the urban, communities had become increasingly disconnected from rural areas. The importance of maintaining vitality in rural villages was not on the public agenda, and the wider society held a rather narrow perspective towards their connection with rural areas. Rural areas were mainly seen as venues for recreational activities, and LCW was considered a local tourism spot by the general population. As such, lacking support from the government and wider society as well as suffering from a loss of population, Indigenous communities had lost the means to drive and guide revitalisation efforts.

The villagers' also possessed strong social capital in their bonds and relations within the village. Many of the Indigenous villagers are related, belonging to one of the five Indigenous families and so the village community possessed close, familial ties. This meant that the villagers were wary of those from outside their community, considering them as 'strangers' that did not understand or respect their community way of life. This is particularly as the villagers tend to define 'home' much more broadly than those from urban areas, perceiving the concept to cover the natural environment and community, not just their place of lodging. As such, this strong social capital had to be harnessed and made malleable through trust building to allow those outside of the village to enter into the community and become productive members.

To solicit support from the wider society, tackling their misconception of rural affairs as private issues was the first step. Part of this required changing private governance structures to public ones, which included and involved wider society. Bringing rural development into public debate, reacquainting urban communities with rural areas became missions that were embedded in the Programme, which later led to public funds being made available for revitalisation projects.

Secondly, the lack of public services, especially transportation due to the village's remote location, made it particularly challenging to introduce more manpower and other resources into the village. Transport to LCW was heavily reliant on the route from Sha Tau Kok Frontier Closed Area, which required obtaining a permit. These permits are not accessible to the public without valid justification and are limited. As a result, only a limited number of people were able to access the site and transport costs were high. LCW's inaccessibility also meant that it was difficult to hire skilled workers such as electricians, plumbers and carpenters. It also meant that there were no schools, clinics or other public services and no entertainment facilities. While the village was able to attract farmers and nature lovers, it also needed to attract new members with other skillsets to sustain and energise the community.

Thirdly, also related to the lack of public services, other than infrastructure the most prominent issues are those of waste management and water pollution. As with many peri-urban areas, increased activities, whether by those living or working in the village or by visitors, brought challenges to protecting the natural environment. While the growing size of the local community offered an important basis for building social capital, it inevitably increased resource use and subsequent waste and pollution. It also created management challenges as common pool resources, which were previously shared and managed by the village community, now had to be shared with others, sometimes for business purposes. This created disagreements amongst the villagers and raised issues of fairness and rights.

In addition, prior to revitalisation, the only means of income generation in the village was related to mass tourism. The villagers operated food stalls, which only marginally contributed to the livelihoods of the local community. Income was also variable, depending on the season and unevenly distributed amongst the villagers. Tourists also bought with them increased litter, pollution and environmental degradation due to poor behaviour as well as paid little heed to local culture. As such, it did not represent a sustainable growth avenue of income for the village. Building a meaningful connection between the wider society and the village was important for the village to be revitalised in a sustainable manner.

As the above challenges were addressed, an overarching approach adopted by the Programme, embedded across many of its initiatives, was to recruit and incubate communities of interest in rural-related affairs. Interested individuals were recruited through a variety of means such as volunteers, apprentice farmers and artists, some of whom are provided with training and opportunities in the village. Through working and/or living in the village, participants not only learn new skills and knowledge, but they also develop a more thorough understanding and build essential connections with the place and the people, which contributed to their acceptance by the Indigenous population. The latter was the essence of the Programme and has led to the desire to become a new settler or part of the village community. The Programme team then linked these new community members with basic resources, such as accommodation and farmlands, and helped them to plan, structure and setup their living in the village.

3.3.3 Building Platforms for Economic Vibrancy and Connectivity

To make the village economically sustainable, its connectivity to the rest of Hong Kong needed to be improved and platforms for economic activities needed to be constructed. The lack of connectivity was a major issue that needed to be addressed as it made many economic avenues unsustainable, due to high transportation costs, and limiting the growth of a new village community. Revitalisation efforts helped demonstrate a need for improved accessibility to LCW and the Programme team and partner organisations, particularly HKCF, supported the villagers request for a new ferry route from Ma Liu Shui to LCW on Sundays and public holidays, which began operation in 2016. The Programme also arranged and paid for a speed boat service, which was made available to individuals working in the village on a daily basis. This significantly improved accessibility to the village and increased the feasibility of attracting skilled people to integrate with the village community as volunteers, partners or settlers.

To date, farmers, artists, architects, surveyors, an electrician, plumber and carpenter, meditation and yoga teachers, photographers/filmmakers, a nature therapist and tour guides have all been involved in the project. While improving accessibility of LCW, protecting this peri-urban village from irrecoverable environmental damage was vital and so avoiding excessive urban influences and pressure on the resources at the village was important. Funding schemes under the Programme were designed to support groups whose work contribute towards safeguarding the cultural and natural characteristics of the village. Those that join the LCW community all possess strong perseverance and demonstrate a passion and willingness to contribute to the sustainable future of the village.

The village's long agricultural history has contributed to efforts to establish sustainable farming in the village throughout the past nine years and progressively more farmland has been recovered. With the intent of attracting new settlers and building the capacity of all interested community members, "3 Dous" Community Building Scheme-a farm start-up initiative, 'farm apprentice' and 'farm volunteer' schemes have been implemented as part of the revitalisation programme. Individuals with an interest in sustainable agriculture are recruited to join the LCW farming community with Indigenous villagers. Through the process of receiving training together in organic farming methods and farm machinery repairing, relationships are built between interested individuals and Indigenous villagers. The Programme encourages them to form into small groups and set up smallholder farms, rather than having everyone working on one big farm. Having small farming groups operating alongside each other means that if one or more of the groups collapse for any reason, others can still continue operating, thus creating a more resilient local agricultural system. Currently, 11[5] groups of community farmers and producers have been established and are in operation.

---

[5] This group consists of nine farms, which includes two project farms, as well as a shared kitchen (LoCoKitchen) and the producer group 'very ginger'.

Rebuilding a farming community

- 3 Dous Community Building Scheme: launched in 2015, provides training and capacity building to Indigenous villagers and volunteers to develop their own farming and/or processing start-ups. There are currently nine groups under this scheme practicing self-sufficient lifestyles, agricultural production and farm-based education, resulting in a diversity of farm products and services. The scheme has contributed to the growing-processing-marketing model later developed under Programme.

- Farm Apprenticeship scheme: a practical based training programme that operates alongside a part time job and allows novice farmers to 'earn as they learn'. It includes skills such as daily farm management through to crop science. The apprenticeship period is a year-long with graduates eligible to join the 3 Dous scheme and apply for support in setting up their own farm at or outside LCW. Five farm apprentices joined the scheme, with two of these graduates setting up a new farming group at LCW.

- Farm volunteer and internship schemes: a variety of opportunities are offered to interested individuals to be engaged in farming at LCW under several schemes, e.g. the 'Farming holiday' scheme offers lodgings for volunteers to stay at LCW while working on the farm and is available to the general public, and the Summer internship schemes recruit local students to gain experience by assisting in the daily operation of the farms at LCW.

- Farmers market: established in 2017, the LCW farmers market is held on the first Sunday of every month to showcase and sell the locally grown produce and homemade products of LCW. The market brings together community farmers and producers under the 3 Dous scheme and is held outside the LCW cultural hub. The farmers market is the first platform to be established with a common objective and shared benefits.

Lai Chi Wo Village Rules

The villagers established rules pertaining to the management of farming responsibilities and watercourses, which include the:

(1) Management of waterways and water resources. These rules include cleaning schedules and procedures of the reservoir and water pit, diversions of water flows and water distribution.

(2) Collective farm arrangements: requirements for participation in collective farm works and procedure if a farming team is unable to make the designated day, registering of farming tasks by the farming teams, timeframe for completion of tasks, reports on tasks, schedule of community farm work days.

Since the establishment of the farming groups, the Programme team continued to provide support for the building of social capital. Multiple Whatsapp groups,[6] have been set up to facilitate information exchange and for ease of communication and coordination between the farms, which compliments regular face-to-face meetings held once every two months. These forms of communication have been crucial in continuing to build trust within the farming community, facilitating the shaping of shared values and norms, which have been particularly important since the introduction of farmers from outside the Indigenous village community. As a result of conflicts over the years, the group has, under the facilitation of PSL and HKCF (the official land tenant), addressed and resolved them by developing rules. These rules cover issues such as decision making of shared facilities, collective branding strategies and community events partnerships. Through this process the community's capacity to resolve conflicts is enhanced and a more effective governance system is built for this SES. The formation of shared rules also aids the integration of new farmers into the village, alleviating some of the mistrust felt by the Indigenous villagers towards outsiders, as social capital bonds are developed between all the farmers of the village, creating a new village farming community. Where the management of village commons previously were conducted amongst Indigenous villagers, new farmers are now involved and share such responsibilities. Regular meetings offer opportunities to make adjustments to management practices as circumstances change, which is important in the dynamic peri-urban SES.

Creating a diversified portfolio of local produce and products is important to support a more resilient economy for the village and economic viability for those living or working there. Compared with the sale of fresh produce, processing produce extends the shelf life and increases value, leading to increased profits and more stable income. Therefore, it is generally beneficial to adopt agricultural and processing activities when building a vibrant rural economy. In the case of LCW, it is even more important that attention is paid to developing processed products rather than the direct sale of fresh produce to consumers. This is because, due to the remote location of LCW, the cost of transporting produce and/or products to be sold outside of the village is high. This is particularly problematic with fresh produce, which have a short shelf life, as their value is low and transportation costs lead to significant increases in price. Consumers have many alternative options, such as buying from other local farms closer to urban areas. Processing produce creates more unique products, which builds a stronger brand and means that there are fewer alternatives. The increase in value and shelf life means that the increased transportation cost constitutes less of an issue.

The Programme team with the farming and producer groups have contributed to increasing the variety of crops and products produced at LCW. The groups are encouraged to maintain frequent dialogue and to grow different crops. This reduces direct competition and creates more economic opportunities for the village by offering more

---

[6] Whatsapp groups have been established between: regular residents/farmer groups, community farm groups, Indigenous villagers (including those overseas) and new settlers, the young Indigenous community, tour guides and volunteers and the many different task groups.

choices for consumers (e.g. there is one producer group, called 'Very Ginger Lai Chi Wo', who, aside from ginger products, also produces turmeric skin care products). Meanwhile, the Programme team continues work with the farmers in experimenting with growing new crops based on sustainable farming principles and the development of new products (e.g. Brown sugar ginger tea) as well as the reinvention of traditional products (e.g. Pickled mustard greens—"*Mui Choy*", which are a key ingredient in the famous Hakka dish of Braised Pork Belly).

There are several members of the village community who wish to promote their own 'home-made' brand. In Hong Kong, however, regulations constrain the sale of pre-packed processed food, which must be produced within licensed food factory/restaurant. Proper sewage, water supply and fire safety measures are also required for the license application, which is incompatible to many village houses and involve high investment inputs.

To support these innovations from LCW and other rural communities, a shared kitchen called 'LoCoKITCHEN' was set up in the Frontier Closed Area of Sha Tau Kok (approximately 4 km from LCW in distance, or a 15-min ride on private speed boat) in 2020 as the first local incubation platform for social innovation on local food. Having obtained a couple of food production related licenses and installed the relevant machinery and infrastructure, the shared kitchen intends to serve local farmers, producers and start-ups, nurturing community-based local agriculture and production development and incubating food-based social entrepreneurs. This will further promote the value of maintaining sustainable rural communities to the wider community, helping to build social capital that bridges these communities. Individuals have also been recruited from the local community, and engaged in participating in other activities that align with the food education mission of the kitchen. This also helps to create social capital by linking these neighbouring communities together.

Supporting the formation of processing platforms
- LoCoKitchen initiatives and activities:
  - An incubation scheme is underway to invite local chefs and ordinary citizens to invent new recipes or rediscover traditional family recipes using local produce.
  - Six women from the Northern District and a farm apprentice have formed a processing team to support fresh produce processing in LCW and Sha Tau Kok. They have been offered training on work safety, hygiene, relevant skills and knowledge and employment opportunities at the kitchen.
  - New ways of processing produce, such as freezing ginger or bottling, are being trailed
  - Provides general training on work safety, hygiene and essential skills.
  - Workshop was held to demonstrate traditional Hakka recipes, such as the making of rice dumplings

> – Establishes a physical spatial presence outside of LCW, which helps engage residents of the Sha Tau Kok Frontier Closed Area.
>
> • Farmshare: social enterprise
>
> – Provides venture support for the LoCoKitchen incubation platform.

The Programme has also facilitated the formation of a new social enterprise, FarmShare, to provide venture support for the incubation platform. Several existing or common approaches were originally considered, such as local Co-ops under the Federation of Vegetable Marketing Co-Operative Societies or the formation of a private company amongst villagers and farmers to hold assets such as the processing facilities and the brand(s) that will eventually be built. Eventually, the formation of a social enterprise to support value-added agricultural activities, a social innovation in itself, was proposed. This represents a new model that could be feasible for the revitalisation of other rural areas in Hong Kong or other regions.

Partnering with a local agricultural advocacy organisation, 'Kong Yeah', and collaborated with food wholesaler, supermarket chain 'Yata', the Programme has diversified the sales channel for LCW produce and products. A regular farmers' market was also established in LCW in 2017 to sell and promote LCW agricultural produce. Alongside popup promotional events in urban areas, this helped to link the rural farmers with customers and strengthen the urban/rural relationship.

3.3.4 Restoring Cultural Assets and Local Capacities Through Social Innovations

With the declining population, the village was at risk of losing traditions, customs, knowledge and craftsmanship. The younger generations were living abroad or in urban centres in Hong Kong and had little interest in rural life, leaving traditions to die with the older generations. This potential loss of tradition and culture would significantly weaken the social capital of the village as it serves to underline the sense of community and social hierarchy that bind the villagers. It also weakened the human-nature relationship built on Indigenous knowledge, which is fundamental for understanding sustainable living within the area's ecosystem and rural context. There was also the potential for such a loss of culture to undermine the overall stability of the village due to a loss of identity and ancestral connections.

Therefore, the safeguarding and the promotion of the value of the rich Hakka history and culture is an important aspect of the revitalisation of LCW. It was important that while modernising and adapting to the modern world, the village's cultural heritage and identity was not lost. Research was conducted into traditional knowledge and practices and conserved through different means such as the documentation of oral history and the reconstruction of a dilapidated village complex. The reconstruction used traditional Hakka construction techniques and locally-sourced materials, turning it into the "Lai Chi Wo Cultural Hub". The Programme team encouraged the Indigenous villagers to conduct knowledge exchange with external knowledge holders such as architects and artists. The Hub serves as an exhibition of the history of the Hakka people's traditional rice farming—displaying farming tools and illustrations of farming methods.

Restored village complex

Under this project, a row of dilapidated village structures and their front yards were restored using traditional Chinese house building techniques and local materials to emulate the original building methods of LCW village. This was done through a collaboration between local villagers, architectural conservationists, academia, builders and specialists in traditional construction knowledge and volunteers. The complex is now a multipurpose shared space, used for Programme information displays, thematic exhibitions, classroom activities and community gatherings. Currently, it is being used as the "Lai Chi Wo Cultural Hub".

The Lai Chi Wo Cultural Hub is leased through the HKCF from its village owner under a time share arrangement. Under this agreement, HKU provided the funding for the restoration of the building, originally a pig shed, in return, they are allowed to lease the building from the owner for a nominal fee of 1 HKD per year for five years. If neither party proposes any changes to be made, the lease would be renewed for another five years under the specified terms. For the initial two years, the Programme has 90% use of the Hub, as the years progress, the percentage of time they can use the space for is gradually reduced to 50% from the 6th to 10th year. This was the first of this type of arrangement trialled by the project as a social innovation, which helps to address issues relating to the spending of public funding on private property. Proving to be successful in LCW, such sharing arrangements have since been applied elsewhere in the village revitalisation context. Under another project at LCW, the HKCF has undertaken to restore village houses in return for a 20 year lease. Under these leases, the owners can use the property exclusively for two weeks of the year initially and this time will then increase throughout the duration of the agreement. A similar arrangement is also present in the neighbouring village Mui Tsz Lam, where two houses are being leased for restoration work.

Under the 'Co-creation of the Community' scheme, the Programme provides funding for talents to initiate projects that engage with local villagers to co-create innovative ways of safeguarding art and natural and cultural capitals at rural communities and build awareness among the public. As the project groups interact with the local community, immerse in the natural environment and history of LCW, they develop unique activities that celebrates the values of these resources and traditions with the local community and promote these values to the general public.

'Co-creation of the Community' scheme

Themes—Hakka Reinvention, Rural Art and Education, Natural Craftmanship, Rural Appropriate Technology and Design (four batches between 2018 and 2019).

Projects:

"Murmur of the brick—Rurally engaged art"

The project transforms a village house into an interactive art installation using mud, hay, earth and adobe bricks, to showcase and commemorate the stories and songs of LCW village. The project also uses traditional and innovative hand-woven fabric to inspire the contemplation of Hakka culture, vernacular aesthetics, traditional crafts and the values of rural villages in contemporary society. The team of artists work with the local community to develop an interactive exhibition platform that uses an art-led participation process to re-examine the cultural significance of past generations. The project also conducts oral history interviews and documentation, desktop research, collects Hakka songs and holds workshops on adobe bricks, Hakka hand-woven ribbon belts and farm revitalisation.

"On Earth—Lai Chi Wo Art Project"

A group of artists developed a series of ceramic and mixed media pieces inspired by LCW's natural environment as well as the stories and memories recovered from overseas Indigenous villagers. An exhibition was organised to showcase this series of work in an urban art centre and in LCW, which helped to generate public interest in the natural and cultural capitals embedded in rural communities as well as helped younger generations of Indigenous villagers to gain a greater sense of appreciation for their ancestral ties. A studio was set up in LCW where workshops were held with LCW villagers and visitors to create ceramic pieces inspired by the LCW nature and culture.

"Nature, Earth and Human—Mui Tsz Lam Art Revitalization Project"

The third project under the scheme, launched in November 2019, revitalised a village house in Mui Tsz Lam, a village neighbouring LCW. The house was transformed into the Mui Tsz Lam Story Museum to hold art exhibitions, showcase Hakka living, sell traditional products and provides a resting place for hikers. Under the project, villagers and artists co-designed murals with the theme "Nature, Earth and Human" to present the story and rural character of Mui Tsz Lam. These murals were painted on the façade of the old village houses by students and volunteers. Workshops and guided tours about the village's environment and culture are also regularly organised.

"The Common Map"

The project creates maps for the village commons. The Common Map is a project that applies cultural mapping as a tool to record, organise and display the local stories, natural resources and farm products of the areas villages and integrate them into an online database for rural capital. The process brings together local and overseas villagers, farmers, students and the general public. Villagers are invited to share their memories and future aspiration of the village, working groups to share their experiences of living and working in the village and farmers to introduce their farms. The general public and students are invited to workshops to sketch the villagers' memories, draw farm maps and collect the stories and aspirations of the villagers. This builds appreciation of the Hakka culture and sustainable lifestyles.

The Programme also organised large-scale events, such as village festivals, which provide another platform for the partners of the 'Co-creation of the Community' scheme to showcase their work to a wider audience and for Indigenous villagers to experience the vibrancy the scheme brings to LCW. Social media coverage of the village festival and newsletters all contribute to rebranding traditional knowledge and capitals embedded in rural communities to raise awareness and interest amongst the wider public.

As the village became more populous and environmentally conscious, new rules were established for waste disposal and innovative recycling solutions were implemented. Traditionally, the village burnt its trash, however, under the new rules waste cannot be burnt within the village area and only wood can be burnt, no plastics. The villagers also passed a rule to ban dumping of construction waste within the village, now waste must be transported to proper government dumping facilities outside LCW. Issues remain, however, as revitalisation of neighbouring villages has increased the traffic and waste being transported through LCW, which as the largest village in the area and closest to the pier, acts as a hub or transit location for its smaller neighbours to place their rubbish before it is collected by the government. In response, the local community has requested an official waste collection point be established, however, the government has yet to determine a suitable location.

Solutions for waste recycling

In response to request from the local community and in a bid to improve resource use, the government established recycling collection points, however, the material is often not properly separated resulting in recyclable material being mixed with non-recyclable waste. Consequently, the waste is disposed as general trash rather than recycled. This is a problem widespread in Hong Kong, not just in rural areas.

Due to this, the villagers have taken matters into their own hands and alternative measures were found to recycle waste, these include:

- Farmers recycling plastic bottles to make tools and cardboard for mulching. Specifically, the community farms have setup their own food waste collection points within the village to generate compost for farming. They collect food waste from neighbouring restaurants within LCW, household food waste from non-resident community members and that from restaurants in Sha Tau Kok.
- Newcomers to the village and artists have upcycled and reused construction waste, for example, the project 'On Earth' used broken tiles to create artwork and bamboo is used for making bio charcoal.
- There has also been a reduction in single-use utensils by some local food stalls, however, more could be done in this area.

3.3.5 The Role of Bridging Organisation and the Incubation of Collaborative Socio-economic Models

Under the Programme, different incubation opportunities are incorporated to mobilise change agents from the wider community to act towards rural sustainability. This is done by providing resources and support for the incubation of social innovative approaches to addressing rural problems as well as the implementation of these innovations. The experience of implementing these initiatives developed through the incubation process could inform revitalisation efforts in other rural areas. In addition to the Programme's support of social innovation, a variation of incubation methods are developed and their impacts assessed. What is common amongst the incubation methods is that they remain community-based (bringing together community of place and community of interests), contribute towards sustainability, help to empower individuals to work collaboratively and sustain the impact.

While the 'Co-creation of the Community' scheme is developed with a focus on developing new ways of safeguarding and promoting natural and rural capitals for wider appreciation, the 'Rural in Action Start-up Scheme' nurtures the development of diversified business models for rural vibrancy. In line with the SES framework, the criteria of these start-up models are that they must be environmental consciousness, financially sustainable, culturally appropriate and socially beneficial (to the local community and/or wider society). During the Programme period, a total of ten projects, proposed by small groups of innovators with a variety of interests and expertise, were funded. Guided by the vision of building sustainable models for the peri-urban context, funding is provided to successful projects that are implemented in any rural area in Hong Kong, so that the networks, resources and impact of the Programme can be extended beyond LCW. Facilitating the building of sustainable SESs in multiple peri-urban communities and to help form a network between them is important for the resilience of these communities.

Rural in Action Start-up Scheme Projects

Cohort 1—awarded in 2019

List of projects

"Cook.Book in Nam Chung"

The project empowers and engages a group of women to lead experiential cooking activities where parent–child duos are recruited to prepare food and enjoy a reading experience in Nam Chung (a rural area in Sha Tau Kok). It aims to build a new business that utilises locally produced farm products, attract urban families to develop human-nature relationships and an appreciation of the natural and cultural capitals embedded in rural communities.

"Bright Bird Biodynamics"

Based in Pat Heung, Yuen Long, a farming team has been adopting biodynamic farming by operating in a self-sustainable model without importing any fertiliser (artificial or organic). The team launched a Community-Supported Agriculture ("CSA") scheme, which aims to recruit community members with an interest in learning and participating in farming. It connects farmers who produce the food with the community who consume the food. The team believes that cooperation under the CSA encourages society to support sustainable food production.

"Pu Giong Zii"

Named after the native plant of the area, 'pu giong', this is the only project in this cohort based in LCW and the adjacent village, Mui Tsz Lam. After learning about the historical significance of this native species to the Indigenous villagers, the project proponents collaborate with villagers to engage in processing and developing new products, such as soap and incense, using pu giong to sell. Villagers traditionally treasured pu giong for its medicinal properties and its symbolic presence, which is understood to represent the perseverance of the Hakka people. In the modern era, however, it is seen as nothing more than a weed. Rediscovering the value of this plant not only helps to develop a possible business opportunity but also helps maintain the village and to foster and reinforce Indigenous villagers' sense of belonging and identity.

"Au Law Organic Commons"

An online platform is created for the sale and delivery of produce and/or products between a network of local organic farms. It aims to help increase income for local farmers by reducing their reliance on limited sales channels and reduce the cost and demand on resources needed to maintain individual platforms and delivery.

"Bring Back Earth Plaster Wall"

The team involved in this project aim to transform people's views on the values of natural materials and to revitalise the skills and traditional building techniques that were once common in the area. This is done by developing an earth plaster using local materials, such as mud, sand, rice straw, oyster shell powder and lime, which can be applied to the walls of homes as an alternative to chemical emitting wall paints.

Rural in Action Start-up Scheme Projects

Cohort 2—awarded in 2021

List of projects:

"A Pearl Treasure"
The project will revitalise old and idle fishing rafts in Sam Mun Tsai fishing village using innovative technology for sustainable seafood production, ecological education programmes and cultural experiential activities.

"Fruitable Hong Kong"
Launch a brand to raise awareness for locally grown fruits and strengthen bonds between fruit farmer communities and the wider Hong Kong community through sharing culture and stories.

"Circular Urban Mushroom Farm"
By recycling used coffee grounds, this project will produce edible mushrooms in urban areas. The spent mushroom substrate will then be used as compost for local farms.

"Upcycled Scent Project"
Scented products are created from locally farmed materials to promote community development through artistic expression.

"Hong Kong Indigo Re-Cultivation Project"
The project will rehabilitate indigo plant species in Hong Kong with the aim of establishing a local six-level industrial chain for indigo dyeing, which covers production, processing and services.

Funding support alone is insufficient. Hence, the Start-up Scheme provides facilitation in a number of ways, including helping participants to connect with local stakeholders and to access local knowledge. Other support is also provided in the form of advice on business set up and ways to enhance their contributions to sustainability issues, promotion through the Programme website and social media platforms. These enable participants to further develop their project design and facilitate them to achieve greater impact.

Interested parties or individuals who need support in developing a proposal for a start-up can participate in one of the Programme's Sustainability Hackathons. The 'Sustainability Hackathon' offers a platform where passionate individuals are gathered and encouraged to collaborate to develop innovative ideas related to the chosen theme. Through the hackathon process participants are provided with a basic understanding of problems relevant to rural communities, such as problems in the local food system leading to its unsustainability, they are guided through a process of brainstorming, idea development, peer learning and coaching by mentors and then they develop a proposal. The Hackathon also acts to fill some urban dweller's gaps in understanding of rural issues as well as providing an opportunity for participants to test the acceptability of their ideas and build their networks in the peri-urban areas. The winners of the Hackathon are encouraged to apply for the 'Rural in Action Start-up Scheme'.

3.3.6 Lessons and Conclusions

Social capital and social innovation are key requirements for building rural resilience. The migration of villagers and the remote location of LCW meant that, prior to revitalisation, social capital was at risk of rapid deterioration. PSL utilised its connections with the Indigenous villagers, skilled individuals and the wider community to bring sustainable living back to the village.

The village had to be sustainable in the modern world. This necessitated for it to be economically viable and to link with the wider urban community, while maintaining its cultural and natural distinctiveness. To this end, a range of economic and co-creation schemes were implemented to harness the social innovativeness of the villagers and the wider community. PSL provides support throughout the application and the implementation of co-creation projects and start-ups. By having a range of initiatives, the Programme created diversified incomes, networks and exchange opportunities for the village. Agricultural production was established to ensure nature conservation and economically stable collaboration, rather than competition, through the smallholder farm approach. The creation of a sharing kitchen and farmers markets links the rural community with the urban. Apprentice schemes and events such as the Hackathon also brought in new individuals with fresh ideas to the community. In this way, PSL was able to harness social innovation to bridge the urban/rural divide and better integrate LCW with the wider community.

To avoid too much adaptation to the modern and urban world, the integrity of the village needed to be maintained and so it was essential to respect and protect the Hakka culture and traditions. This was essential in building social capital amongst the Indigenous villagers and new community members as it provided a sense of belonging and identity. PSL provided leadership in documenting traditional practices and the villagers' history and in organising restoration activities to bring the community together and foster a sense of ownership for the village. It was also evident that more people were required to create a community from which social capital could be generated. PSL's leadership in recruiting and integrating newcomers into the village to build such a community was found to be essential for the Programme's success.

To build the village community, interested, socially and environmentally responsible parties are brought to LCW and introduced to villagers, successful applicants are then facilitated with renting a house in the village. The building of trust between PSL and the villagers is crucial for PSL's role as the bridging organisation in introducing newcomers into the village. Indigenous villagers who do not live in the village usually leave their ancestral homes unoccupied as the monetary return for renting is unattractive and they are generally reluctant to rent to outsiders. This is where the role of PSL and the relationship it has built with the village becomes critical, they recommend their funding scheme awardees to Indigenous villagers and help liaise between the two for a lease. Through living in LCW, these newcomers develop a sense of attachment to the village and villagers, building trust, reciprocity and collaborative relationships. This, again, widens the social networks and the ability of the local community to mobilise resources to support the village in the face of stresses and shocks.

As a result of these processes, LCW has become a place rich in natural, cultural and community assets. Revitalisation efforts empower the community's capacity to reinvent its uses of rural assets for sustainable development. This includes developing new socio-economic models, such as community farming, farm-to-table activities, Hakka cultural experiential events, nature art and handicrafts, to enhance the capabilities and competitiveness of the local community.

3.4 Governance: Towards Polycentric Institutional Arrangements

3.4.1 Introduction: Polycentricity and Participatory Governance

Polycentric institutional arrangements better equip rural communities in managing shocks and unexpected events. This is particularly important to those in the peri-urban context as they are prone to not only changes in the natural environment but also highly complex and dynamic urban influences. The ability of effective learning and adaptation are enhanced through polycentric arrangements (Berkes & Folke, 1998) and so local governance is better equipped to the SES in which it operates. Such arrangements also enable increased level of participation. Participatory decision modes that facilitate collective learning are considered a prerequisite for the advancement of sustainable policies (Dryzek, 1997), especially when there is often an implementation deficit in environmental policy (Knill & Lenschow, 2000) and societal structures are increasingly complex (Fritsch & Newig, 2012). As such, it brings the local level into the governing regime, which polycentric institutions can then link with higher authorities.

Participatory governance principles were incorporated into the Programme's structures and processes. Participatory modes of governance provide the means for environmental goals to be achieved in a more targeted, swift and effective manner (Bulkeley & Mol, 2003). The inclusion of a range of stakeholders and citizen groups serves to diffuse information, allows for consultation and supports sharing in anticipation of the future as well as supports the coordination of different forms and fields of knowledge, the coproduction of solutions and social learning (Brousseau & Dedeurwaerdere, 2020). Participation also enhances learning processes, improves the quality of decisions and can contribute to empowerment, which in turn contributes to enhancing the resilience of the system. Qualities such as transparency, equity and accountability in decision making processes are difficult to achieve but are important to strive for to achieve effective self-organisation. In the LCW situation, collective identity also had to be enhanced to increase the willingness of the villagers to participate in governance processes.

Similarly, the Programme team took a leadership role in building the capacity and motivation of both individuals and the community to apply sustainable management in collective decision making. The institutional situation also needed to be reformed to a more polycentric arrangement, with governance structures altered to reflect greater participation and institutional diversity cultivated. Institutional diversity also needed to be increased so a range of institutions were created under the Programme. This includes the creation of sub-programmes and initiatives to handle the different elements of the Programme. These institutions took different forms, from incubation schemes to workshops and community committees. A diversity of management approaches enshrined in the different institutions enables learning and understandings of the optimal approaches to managing SESs.

3.4.2 Principles of Good Governance and Collective Identity

Engaging the community and different groups provides a diversity of perspectives, which aids the understanding of SES dynamics and enhances the resilience of the SES (Biggs et al., 2012). This was incorporated in revitalisation efforts through participatory approaches, which ensured that transparency, equity and accountability in the decision making and running of these institutions were maintained throughout the Programme's activities. Transparency, equity and accountability are recognised principles of good governance (UNESCAP) and are also cornerstones of true participatory governance.

Transparency requires information to be freely available, accessible and understandable to those affected by decisions. It requires clear processes and procedures and so ensures the accountability of the governance system. Without this openness, there is the risk of administrative corruption (Kim et al. 2005). In addition, to ensure the equity of decision making, participation has to be inclusive. This meant collaborating with both the villagers residing at the village, those present elsewhere in Hong Kong and those abroad, as well as bringing in new settlers in an inclusive manner.

The isolated nature of LCW and the bureaucratic nature of the Hong Kong govern-ment, coupled with mistrust felt by the villagers towards those they considered to be 'outsiders', meant that transparency and accountability were difficult to achieve in this peri-urban situation. Power imbalances between the government and the villagers also needed to be addressed to ensure the equity of the Programme and that the villagers' rights and culture were adequately protected and preserved alongside the natural environment and ecological welfare of the area. The Programme team created opportunities where the villagers, those residing locally and abroad, government and concerned citizens could be represented in decision making within the revitalisation process.

Ensuring transparency and equity through a wider community focus

The LCW Programme is not alone in its revitalisation endeavours at the village. After its first phase, several partner organisations and separate programmes have since begun operating alongside the LCW Programme. PSL's approach somewhat differs from others and has placed a heavier emphasis on engaging the wider community and implementing revitalisation initiatives beyond the village. This is due to its focus not just on the Indigenous village community or the local LCW community but its consideration of connecting LCW with the wider community and it's vision of developing sustainable models for other villages in the peri-urban context.

As the revitalisation Programme is publicly funded, PSL seriously considers how to balance the needs and demands of the whole society, resulting in a more broadly accountable approach. Some of the other publicly funded revitalisation projects struggle to find a balance between a top-down and bottom-up approach to revitalisation and so issues arise from the competing needs of society and the villagers. The result is often promising too much to the villagers, furnishing them with unrealistic expectations, and a subsequent loss of support from villagers and the public for the project.

To cultivate trust and mutual understanding between all the parties, an open, trans-parent and accountable process that helps to build mutual understanding was required. Firstly, key Indigenous members/leaders who shared a similar vision regarding rural revitalisation and the development of the village were identified and became partners. One of the HKU team members is an Indigenous villager in another village in the area, as both an Indigenous villager and an ecologist he was in a propitious position to bridge the divide between Indigenous villagers and civil society, gradually building trust between the groups. Another prominent figure was Lam Chiu Ying, a former Director of the Hong Kong Observatory. Due to his position as a public figure and scientific reputation, he played an influential role in trust building in the initial stages of the Programme.

Alongside these principles of good governance, collective identity is important in motivating local community members to engage in collaborative decision making and mobilising collective responses to socio-ecological challenges. Collective identities inform if and how individuals work cooperatively across various differences to develop an adaptive response to SES transformation and so impact the resilience of the system (Leap & Thompson, 2018). Collective identities are present when there is a shared, positive sense of self across a group of individuals, which aids the alignment of individuals' perceptions regarding who they are and how to respond to a situation (Polletta & Jasper, 2001). This also enables individuals to respond collectively to challenges through a shared perception of a problem and so the solutions to these problems. As such, the trust and respect that originates from collective identities facilitate individuals working together to build resilience in response to changes in SESs (Adger, 2003; Leap & Thompson, 2018). It is also important in building a sense of ownership and a personal feeling of responsibility towards sustainability.

While the older generation of Indigenous villagers maintained a strong collective identity through their shared cultural heritage and history, the younger generation were much more detached from village affairs and so had little involvement in the revitalisation project. Many of this generation were born overseas and so had a weak connection to Hong Kong and those that remained were preoccupied with work and urban life. The handling of village affairs is also traditionally managed by village seniors, which further distanced the younger generation. As a result, the succession of traditional rituals and customs is weakened and collective identity was eroded. The collective identity of the LCW villagers, therefore, needed strengthening to ensure both their continued participation in the governance processes and traditional cultural events as well as to reconcile the villagers living locally and abroad and the younger generation with the older one to build a coherent vision and a sustainable future for the village.

Safeguarding cultural identity

The village's cultural identity was protected and nurtured through a holistic approach. The implementation of project activities related to eco-agriculture, co-creation and start-up schemes all help to re-establish human-nature relationships, safeguard Indigenous knowledge and skills. These are then promoted to the public through various channels, the major ones include the Academy, the annual village festivals, art events/exhibitions that occur in or outside of the village.

Further details:
- Co-creation of the community scheme projects, such as 'Murmur of the brick' (see Sect. 3.3), which also led the first two days of 'Lai Chi Wo! Village Fest 2019' with the theme 'art for all'. Here, a Hakka group performed mountain songs and knitted Hakka embroidered bands and clothes. Aside from promoting Hakka culture, it aimed to reignite the villagers' sense of attachment and appreciation of their cultural identity.

- The Academy for Sustainable Communities (the Academy) is housed under the Centre for Civil Society and Governance at HKU. It offers curriculum-based courses, seminars, forums and field-based activities covering sustainability-related knowledge and skill sets. Its vision is to become a regional knowledge exchange platform to disseminate knowledge of sustainability and incubate a new generation of change agents for sustainability.

Other channels to promote human-nature relationship and Indigenous knowledge and skills:

- The Programme team organises or assist in the organisation of numerous traditional Hakka cooking or demonstration sessions with the general public in urban Hong Kong and with visitors to LCW. Dishes include steamed glutinous rice cakes and pickled mustard greens and these sessions occur on a regular basis.
- An exhibition is held at the Lai Chi Wo Cultural Hub, documenting the Hakka people's traditional rice farming methods.

Revitalisation efforts, therefore, had to find a suitable entry point to motivate the participation of the younger generations. The LCW Programme instigated updates and news about the project through mediums such as 'Whatsapp', social media and a bilingual newsletter to increase interest and instigate a sense of belonging. The Programme also had success in engaging them in project activities and workshops, including traditional festivals, art installations and ceramic workshops. Unlike the older villagers, the younger generations are also more open to new ideas and supportive of their family homes being utilised in an innovative manner. For example, one such villager is an architect, the "Murmur of the Brick" project curator invited him to re-invent his family house into an exhibition to showcase his children's memories about LCW and the house.

The collective identity of the village has been enhanced through the revitalisation process. Meanwhile, the social capacity and motivation of the community to apply sustainability principles in collective decision making were also observed. Young villagers were also involved in the decision making process while the cultural heritage of the village was maintained, so as not to alienate the older generations or lose the rich Hakka culture. This has increased motivation and the desire to see the village as a functioning and self-sustaining system, relevant to the modern world.

3.4.3 Village Governance and Institutional Structures

A diversity of institutions provides the basis for learning, innovation and adaptation to continuous changes in the SES. Having a diversity of institutions means that they will respond to changes and disturbances in the SES differently. A variety of organisational forms that possess overlapping domains of authority and operating at different levels allows for a diversity of responses that enable the maintenance of the SES when faced with disturbances (Biggs et al., 2012). Although the presence of a multitude of institutions poses significant challenges to coordination and conflict resolution if overlapping domains of authority are not handled with care.

The LCW village lacked a structured governance system and the village heads faced a legitimacy crisis. While a few Indigenous villagers were sometimes involved informally in decision making, the majority of the time, the two village heads were the only decision makers on behalf of the entire village. The decision making process lacked transparency and was subjected to vested interests especially in the Programme's initial stage. Consequently, village politics often became heated, with families fighting for control of resources. There were also issues with communication as villagers outside Hong Kong did not always receive the correct information, exacerbating tensions. The physical infrastructure of the village also presented challenges as farms are interlinked by shared infrastructure. This created management and governance challenges over issues such as the maintenance of irrigation channels, solar powered electric fences and the weeding of shared farmland bunds, which often resulted in disputes.

As a result of revitalisation efforts, the traditional village management committee, 'Pui Shing Tong', was revived. The committee involves members of four of the five[7] Indigenous families of LCW and makes decisions regarding village affairs. An annual village meeting usually occurs every October, which aligns with the return of many villagers for autumn ancestral worship. This meeting enables villagers to discuss village affairs, resolve major disputes and make communal decisions. Also in October, an annual community meeting is co-organised by the Programme and HKCF, although it was unable to go ahead in 2020 due to the Covid-19 pandemic. While its functions are limited, the meetings allow the team to discuss project developments with overseas villagers as well as engage all the villagers in discussing the sustainability of the village. The Programme has also proposed that the village innovate its organisation system by establishing a shared company amongst the villagers to manage communal income and expenses, although this venture has encountered legal difficulties and so progress has stalled.[8]

From the onset, it is apparent that there is the need for an institution to take over running efforts from institutions such as PSL and handle the future governance of the village. After discussion, one Indigenous villager proposed establishing a social enterprise, 'HakkaHome@LCW', to contribute to the governing and management of the village. The board was initially formed of three Indigenous villagers and two directors of HKCF, who joined as independent representatives of society. One of the

---

[7] The fifth family chose to not be involved in the committee due to its strong opposition to the other current village heads. Underlying village politics and the HKCF's 'Village house adaptive reuse' project sparked the conflict due to a failure of communication with the village heads and other villagers.

[8] Under Hong Kong law, non-profit companies are unable to be registered. This meant that the village company would have to be registered as a charitable organization, which raised concern amongst several of the villagers.

villagers later resigned her directorship to take up a consulting role to the enterprise. The intention was for the enterprise to be non-profit, channelling money back into the village but this has yet to be apparent. Initially, there was little interest from the villagers in the enterprise, hence their limited involvement. As one villager put it, no functional "structure can exist if there are only bosses but no staff", the enterprise had a governing body but no management or workers. This meant that the enterprise had little recognition within the village and was not perceived to be representative of the villagers.

This became problematic when the HKCF, with the funding support from Hong Kong Jockey Club Charities Trust, named HakkaHome@LCW as a partner in a project to restore village houses for experiential learning and tourist accommodation. The social enterprise was intended to take over the management of the restored houses, however, the villagers do not consider the management or running of the enterprise as a priority. The active management of these restored houses would require bringing in dedicated staff. Despite the village's peri-urban context, its relative remoteness compared to urban Hong Kong and lack of structures results in difficulties in hiring staff to support the work of HakkaHome@LCW, hence it lacks the institutional capacity to run the project.

The difficulties encountered establishing and legitimising a social enterprise to govern village affairs demonstrates the challenges of modernising traditional governance structures. Consequently, strengthening and reviving more traditional governance platforms, such as the village management committee, was found to be more practical in engaging villagers in certain governance issues in their village. To illustrate this, Pui Shing Tong assumed the role of a middle-man for sub-letting farmland in the village. With the fragmentation of land ownership at LCW being further complicated by many of the landowners being overseas, significant problems were created in relation to renting farmland. There were also instances where landowners had died but the official ownership transfer procedure had not occurred due to the desire to avoid tax. While the Programme originally facilitated villagers to form a company, 'Ying Wo', to handle these issues, similar to HakkaHome@LCW, the company was subjected to issues of legitimacy and so, a lack of trust from the villagers. This was later addressed by Pui Shing Tong taking over the role and landowners now work under the framework of this traditional village governance institution, renting farmland to external parties.

Communication with local and overseas villagers has been enhanced through the establishment of a 'Whatsapp' group to encourage greater participation and involvement. This platform allows the village heads to announce news to local and overseas villagers, ensuring information is received by all in a uniform, timely and accurate manner. It serves to engage villagers all over the world in the revitalisation of their village and ensures those villagers not physically present are still represented and have the opportunity to voice their opinions and concerns. This platform, however, is only sufficient for distributing information and the exchange of opinions. As such, it has not been effective in conflict resolution, which points to the need for more diversified institutions for improved village governance.

The introduction of new settlers to the village had to be handled with care to ensure equity as well as avoid conflicts. Particularly as the strong social capital bonds between the Indigenous villagers made them wary of 'outsiders'. The LCW Programme acted to bridge the divide between villagers and outside actors and facilitated trust building by encouraging villagers and outsiders to work together on activities such as farmland preparation, rice planting and village cleaning. These activities inspired community cohesion and a sense of belonging, gradually resulting in the formation of a new community of villagers, settlers and volunteers. The villagers also set three rules for new settlers to abide by to ensure that their culture and heritage were appropriately respected and the equity of the project. These were that new settlers had to respect the Indigenous culture and community, would not promote foreign religions and would avoid direct business competition with existing Indigenous villagers. In terms of physical agricultural infrastructure, a facility sharing community was formed amongst the farmers where they could strengthen their mutual support. The majority of farms share responsibilities to maintain infrastructure and equipment and so members are encouraged to take collaborative actions in problem solving. There are also regularly scheduled 'day of community farm work' to pool manpower from every farm to undertake essential maintenance works.

Community farmer meetings are held regularly as a decision making platform. These meetings are hosted by the land tenant and the manager of the Hong Kong Countryside Foundation with a member from each farm. These meetings serve to provide farming news, follow up on maintenance work, formulate and discuss community rules, plan collaborative marketing and promotion events, resolve disputes, share and exchange resources as well as to identify potential risks and discuss preventative and mitigation measures. The revitalisation efforts encourage voluntary support and the exchange of manpower within the community. The co-management approach and communal nature of the farmers' community cultivated under the LCW Programme is evocative of the village's traditional management system.

3.4.4 Cross-Sector Collaboration

Cross-spatial and sectoral partnerships help to enhance the resilience of rural communities. Those in the peri-urban context are better positioned for such partnerships to be forged. As such, many components in the LCW Programme are designed to encourage the involvement of a wide range of individuals and organisations from different backgrounds. As rural communities become better connected with other SESs and sizable communities of interest, the resources that become available to them and potential solutions that can be developed to address any shocks and stresses increases exponentially. This means that the engagement of younger generations of Indigenous villagers and recruiting others to contribute to and/or become part of the community are integral to the resilience of village communities. The bridging organisation's role in nurturing communities of interests regarding aspects related to sustaining a rural village becomes highly important. In this case, the Programme team was heavily involved in (i) addressing a pre-existing condition that prevented cross-sectoral partnerships, namely preconceptions and a lack of trust between villagers and the larger community and (ii) building new cross-spatial and sectoral partnerships.

Prior to the involvement of the LCW Programme, retired Indigenous villagers possessed a 'home-coming dream'. This dream inspired efforts to link the Indigenous community with organisations and groups eager to discuss village revitalisation. This process went through several iterations from 2011 to 2013 but to no avail due to differing visions about village development, the Indigenous community favoured a more tourism orientated approach while the other side had a more ecovillage vision for the area. As a result, no concrete action plans were generated. The leadership of PSL as a bridging organisation was vital in mobilising resources and manpower to catalyse a coherent plan of action and reconcile these visions.

Due to the peri-urban nature of the village SES before the revitalisation process could begin in earnest, institutions had to be developed to build trust both with the villagers and between villagers and the larger community. In Hong Kong, Indigenous communities are often mistrustful of outsiders, including green groups, city folk or the government. LCW was no exception, the village has a long history with its own set of rules and customs. There was a lack of trust from the villagers towards the Programme team, partner organisations and the government. This was largely based on the perception of 'interference' from the government and green groups, who some villagers felt may deprive them of their traditional land rights. On the other hand, there is a perception amongst the wider Hong Kong society that villagers are back-wards, selfish and greedy, forming a 'privileged class' due to their right to land (an expensive and rare commodity in Hong Kong). The green groups in particular were very suspicious of any development-type actions undertaken, opposing Programme activities such as the clearing of vegetation to revitalise farmland due to concerns that the environment was being destroyed. This perception was exacerbated by Indige-nous villagers that sold their property to unscrupulous developers (Hopkinson & Lei, 2003) and has resulted in green groups pushing for greater conservation protection in the village areas.

PSL initiated meetings to open dialogue with the villagers and also participated in village festivals and events. The Programme team meets formally with the majority of Indigenous villagers at their annual village meeting when Indigenous villagers residing overseas return for their traditional festival. The Programme team's staff maintain a close relationship with the Indigenous villagers at LCW and the new settlers, interacting with them on almost a daily basis as they live and work alongside each other in the village. This allows the Programme to maintain a visible presence and ensures their accessibility to the villagers. It also allows for rapid responses to any changes in the on-site situation. The Programme also conducts regular meetings with its partner organisations, the HKCF, the Conservancy Association and Project Green Foundation every three to four months to ensure the goals and activities of the different organisations align.

To tackle pre-existing mistrust between the villagers, revitalisation efforts and the wider community, a forum was organised in 2015 to encourage constructive dialogue.

Green groups, local farmers, relevant industries and key Programme partners as well as any interested parties were invited to attend. The forum explained the intentions behind revitalisation efforts and the emphasis on sustainability as well as provided an opportunity for stakeholders to engage in discussions about revitalisation. As a result, concerns about profit-led development were dispelled and increased support from the community was gained.

Additional efforts were made to reassure green groups and gain their trust. Experts in ecology, hydrology and geography are involved in researching the area and monitoring the agricultural impacts on ecologically sensitive areas, such as the mangroves. Engagement meetings were also held with the green groups to present the plans for revitalisation and to provide them with the research results and assessment findings. These efforts were vital in building trust with the green groups and gaining their support for the Programme.

The revitalisation movement consists of a range of sub-programmes and activities designed to bring together the LCW Indigenous community and interested actors to experiment with possible ways to rebuild the community. As they work together, a mutual understanding can be developed allowing for genuine communication and learning. For example, at the start of revitalisation efforts, interested city folks were engaged to work with villagers and Programme staff on farmland preparation, rice planting and village cleaning. Other sub-programmes were also designed to create opportunities for interested individuals to establish a longer-term connection with the village and the villagers, for example the 'Co-creation of the Community' scheme and the 'Rural in Action start-up scheme'. This involved PSL bringing all the actors involved in the Programme together to develop a guiding shared vision for sustaining the village faming landscape, restoring village vibrancy, continuing Hakka culture and protecting the liveable village environment.

In addition to encouraging the flow of people and resources from the urban areas to LCW, revitalisation has also connected processes in the village with those at wider spatial scale. As a result, the village's position along the rural–urban spectrum is shifted. For example, it has developed networks across the local agricultural food systems, building a chain of production—processing—marketing processes that brings together farmers, processors, marketers and customers. These partnerships not only contribute to the economic vibrancy of the village, which is a crucial pillar supporting its resilience in the longer term, but also enhances the development of the local agricultural and processing industry. One of these initiatives is the establishment of the 'LoCoKITCHEN'. The kitchen is fully licenced in compliance with Hong Kong laws and provides a venue for the villagers and farmers in the region to develop and market their own products. Through the kitchen, PSL also provides training in processing local produce, developing new recipes and products and the marketing of these products.

The introduction of coffee farming not only provides evidence to support wider local application of agroforestry, it also facilitate the building of a range of connections across Hong Kong. The endeavour at LCW integrates shade-tolerant coffee trees amongst other agricultural plants to create an agroforestry habitat for conservation purposes. Local roasters were engaged to help experiment and develop a roasting profile for the LCW coffee and local coffee shop brewers and experts were contacted to try the coffee and build its tasting notes and reputation. Local farmers from across Hong Kong were encouraged to form a coffee grower group to allow for collective learning and local coffee educators were engaged to promote localised food systems and the sustainable production concept. In this way, farmers and LCW were connected with coffee brewers, experts and educators in the wider Hong Kong society.

Recognising the importance of nested SESs to the resilience of rural communities, opportunities are created for LCW to become better connected with other villages. The 'Rural in Action Start-up Scheme' funds projects that can be based in any rural community in Hong Kong, expanding the network of rural start-ups from LCW to other villages. Lessons learnt from the implementation of these innovative models can be shared to enhance the sustainable development of not only this Hong Kong network of local rural communities but potentially those in other peri-urban contexts.

3.4.5 Inter-level Coordination

PSL has been able to act as a bridge for the villagers in communicating with the government and garnering their support. For many years, the villages in the area, particularly Mui Tsz Lam, Kap Tong and LCW, had requested a new ferry route from the town of Ma Liu Shui to LCW. The introduction of revitalisation initiatives further demonstrated the real need of the community and general public for improved access to LCW. With the support of the Programme team and the HKCF, the villagers liaised with different government departments and were able to provide evidence for the necessity of the ferry route, which began operating in 2016.

The Programme also acted as a bridge between the government and the villagers on the issue of the government's Home Affairs Department construction of a walkway. The walkway was to be built within the protected marine park, which meant the Agriculture, Fisheries and Conservation Department as well as the villagers needed to be involved. The villagers, however, are not experienced in talking to government departments and the Home Affairs Department does not often engage with NGOs, such as the Programme team. While the Programme team had expressed their concerns to the government about the design and its impact on the areas' ecology, they were unaware of the scale of the development until a typhoon destroyed the barriers revealing the extent of the construction. The contractors for the walkway also attempted to persuade the villagers to sign a letter stating that they agreed with the construction of the walkway, despite the villagers claiming never to have seen the plans for the project.

Using their position as academics at HKU, two prominent members of the Programme wrote a letter to the Home Affairs Office. The Conservancy Association, a partner organisation with the Programme, also held a press conference to draw attention to the issue. As a result, the Home Affairs Department began to engage with the Programme team and an improved design for the walkway was agreed upon. The Programme was able to utilise its role as a bridging organisation to coordinate between and engage the different stakeholders. The incident also changed the dynamic of the village as more people, Indigenous and non-Indigenous villagers, government officials and NGOs, became more involved in village affairs, enhancing the village community.

This case study of this peri-urban interface provides further evidence on issues that can arise from conflicts between formal and informal governance in the management of common pool resources. The management of the village stream remains a challenge due to differing remits and understandings of government departments and the village's needs. Traditionally, the whole stream was managed by the LCW village, however, the introduction of different government departments and changes in designation of the stream has limited the villagers' management remit. Even though inter-level communication has improved, this example demonstrates the difficulties that remain when dealing with common pool resources where a range of bureaucratic government bodies, with overlapping and conflicting jurisdictions have become involved. Here, government authority stood in the way of localised efforts to manage the stream.

Challenges of common pool resource management

In Hong Kong, the Drainage Services Department (DSD) is the principle technical department that manages flooding and streams, while the Agriculture, Fisheries Conservation Department (AFCD) is responsible for ensuring irrigation provisions to local farmers and the Home Affairs Department is responsible for community level projects.

In LCW, the stream is divided into two sections of management, the upper stream is managed by AFCD as it is considered of ecological importance and the lower section is managed by the local District Office under the Home Affairs Department.

Extensive silt has built up since the cessation of agricultural activities and so needs to be cleared. As the upper part of the stream is designated as being ecologically important, machine dredging is prohibited. Manual dredging is allowed but its impact is marginal. The DSD is also reluctant to reinforce the stream banks due to the low village population and AFCD does not possess the required resource.

Silt accumulation and broken stream banks continue to create significant problems for agricultural revitalisation efforts, as it causes problems for irrigation and issues during drought and floods. There are limited actions that can be taken by the farming community due to government control over the upper part of the stream.

> PSL was able to aid collaboration between the villagers and government in the management of the village stream to some extent. As a result, communication and collaboration between the villagers and the various government departments regarding the stream's management did increase. Differences between the two sides remain and a suitable management approach has yet to be found.

The Programme's and community's efforts and success in the revitalisation of LCW caught the Hong Kong government's attention and triggered the establishment of a new institution. In 2017, the government explicitly referenced the Programme in its policy address (Policy address 2017) and announced the creation of the Countryside Conservation Office (CCO). The CCO was established in 2018, it started out with the objective of enhancing countryside revitalisation in LCW and the ecological conservation Sha Lo Tung, with the aim of extending these initiatives to other countryside areas. The Office has conducted onsite inspections in LCW and liaises and coordinates with the relevant government departments as well as NGOs and stakeholders to discuss minor improvement work. The government also made a fund of HKD 1 billion available to the CCO for relevant conservation and revitalisation efforts. This institution essentially provides more direct access to resources for rural communities (EPD).

> **LCW and the Countryside Conservation Office**
>
> The LCW Programme team has formed a close relationship with the new CCO. Being new, the Office lacks staff with sufficient background and expertise in rural issues and so it often looks to the Programme team at HKU to help identify potential issues with projects, many of which, while well intentioned, are run on a more 'trial and error' basis. The Programme team's informal role in advising the Office makes the wealth of knowledge they gained during the LCW Programme available to the government and helps set an example for NGOs and other groups to follow when working with the CCO on rural projects in Hong Kong.

3.4.6 Connectivity to International Platforms

In 2020, LCW won special recognition for sustainable development in the UNESCO Asia–Pacific award for Cultural Heritage Conservation. The award was significant as it acknowledges the significance of conserving cultural heritage in striving for sustainable development. The achievement of the award compounded interest from the HKSAR government in the LCW Programme, as they requested for the

Programme team to share the approach taken during the revitalisation project at LCW.

UNESCO Asia-Pacific award for Cultural Heritage Conservation

The LCW cultural landscape received the UNESCO award for Cultural Heritage Conservation, the application was built on the HSBC Rural Sustainability Programme, based on which the jury applauded the Programme for its pioneering approach to reviving the village, which "transforms notions of heritage practice from its conventional focus on material conservation to encompass living heritage in all its manifestations". The Programme also received recognition for its demonstration of the "importance of interweaving nature and cultural heritage in setting a new urban–rural sustainability agenda".

The achievement has drawn attention to the fact that Hong Kong has a rural landscape, something that is often overlooked in the city's image as an urban metropolis. It illustrates how conservation of peri-urban areas can contribute to the city's sustainability as, through horizontal integration between the Programme and the urban community, the revitalisation of LCW has contributed to wider social sustainability in Hong Kong.

On a wider scale, the recognition from UNESCO has increased overall awareness of how conservation relates to sustainability. In particular, the SES approach undertaken by the LCW Programme demonstrates how conservation is not limited to the architecture or physical infrastructure of a place. Rather, factors such as biodiversity, community and cultural heritage and practices, which make up the 'living system' of the place, are equally important for sustainable development.

3.4.7 Lessons and Conclusions

The governance structure of LCW needed to be revamped and reformed to reflect changes in the village while being respectful of traditional roles and institutions. Traditional governance structures such as the village committee were revived and re-empowered in their handling of village affairs while platforms such as social media were utilised to connect villages across generations and geography. PSL's leadership here was influential in bringing other stakeholders together and forming a more coherent vision of village management.

For this to occur, trust had to be built, both between the villagers and those involved in the revitalisation efforts and between the villagers and the wider Hong Kong society. The Programme team maintains a close relationship with the Indigenous villagers, staff are present and hands on, sharing in village life. Their participation in village festivals and events further broke down barriers between the villagers and the team. With outside organisations, such as green groups, and the wider public the Programme team increased dialogue and discussion, sharing their actions and findings in an open and transparent manner.

PSL was able to utilise its role as a bridging institution to connect the villagers with talents and markets in the wider society. This was done through establishing physical and virtual platforms to sell and promote local products such as LocoKitchen, and by bringing in innovative ideas through the various incubation schemes. PSL also played a valuable role in bridging relations between the villagers and the government, helping to even the playing field and engage the relevant government officials or departments.

When reforming the village governance structure, the villagers were more amenable to familiar governing structures. More innovative approaches, particularly the setting up of different companies or social enterprises, were often met with disinterest or mistrust and suffered with issues of legitimacy. The Programme had greater success in setting up committees or sharing platforms, which often took over the roles of the companies or social enterprises.

It was also found that, while significant progress has been made in the collaborative management of the village's commons, issues remain where higher levels of government authority were involved. The management of the village stream still presents difficulties due to the overlapping bureaucratic nature of the government authorities and stringent regulations. It is hoped, however, that with the improved communication channels and as the village population and production increases a satisfactory solution will be achieved.

References

Adger, N. (2003). Social capital, collective action, and adaptation to climate change. *Economic Geography, 79*, 387–404.

Berkes, F., & Folke, C. (1998). Linking social and ecological systems for resilience and sustainability. In F. Berkes, C. Folke, & J. Colding (Eds.), *Linking social and ecological systems: Management practices and social mechanisms for building resilience* (p. 27). Cambridge University Press.

Biggs, R., Shluter, M., Biggs, D., Bohensky, E. L. Burnsilver, S., Cundill, G., Dakos, V., Daw, T. M., Evans, L. S., Kotschy, K., Leitch, A. M., Meek, C., Quinlan, A., Raudsepp-Hearne, C., Robards, M. D., Schoon, M., Shultz, L. & West, P. C. (2012). Towards principles for enhancing the resilience of ecosystem services. *Annual Review of Environment and Resources, 37*, 421–448

Brousseau, E., & Dedeurwaerdere, T. (2012). Global Public goods: The participatory governance challenges In E. Brousseau, T. Dedeurwaerdere, B. Siebenhüner (Eds.), Reflexive Governance for Global Public Goods. The MIT Press.

Bulkeley, H., & Mol, A. P. J. (2003). Participation and environmental governance: Consensus, ambivalence and debate. *Environmental Values, 12*(2), 143–154.

Dryzek, J. (1997). *The politics of the Earth: Environmental discourses.* Oxford University Press.

Fritsch, O., & Newig, J. (2012). Participatory governance and sustainability: Findings of a meta-analysis of stakeholder involvement in environmental decision making. In E. Brousseau, T. Dedeurwaerdere, B. Siebenhüner (Eds.), *Reflexive Governance for Global Public Goods.* MITPress, pp. 181–204.

Gebre, T. & Gebremedhin, B. (2019). The mutual benefits of promoting rural-urban interdependence through linked ecosystem services. *Global Ecology and Conservation* 20

Grooten, M., & Almond, R. E. A. (Eds). (2018). *Living planet report—2018: Aiming higher*. In M. Grooten, & R. E. A. Almond (Eds.), World Wildlife Fund International, Gland, Switzerland.

Hopkinson, L., & Lei, M. L. M. (2003). *Rethinking the small house policy*. Civic Exchange.

IPBES. (2019). *Summary for policymakers of the global assessment report on biodiversity and ecosystem services*. Intergovernmental Science-Policy Platform for Biodiversity and Ecosystem Services, Bonn, Germany.

Kim, P. S., Halligan, J., Namshin, C., Oh, C. H., & Eikenberry, A. M. (2005). Toward participatory and transparent governance: Report on the sixth global forum on reinventing government: *Public Administration Review, 65*(6), 646–654. https://doi.org/10.1111/j.1540-6210.2005.00494.x.

King, R. (2018). *Interventions: Sustainable agriculture production*. Hoffmann Centre for Sustainable Resource Economy, Chatham House

Knill, C., & Lenschow, A. (Eds.). (2000). *Implementing EU environmental policy: New directions and old problems*. Manchester University Press.

Lam, W. F., Law, W. W. Y., Yau, W. Y., & Yiu, S. Y. (2018). *Introduction to cultural landscape management, Hubert project, policy for sustainability lab, centre for civil society and governance*. The University of Hong Kong. https://hubertproject.org/hubert-material/455/. Accessed June 24, 2021.

Leap, B., & Thompson, D. (2018) Social solidarity, collective identity, resilient communities: Two case studies from the rural U.S. and Uruguay. *Social Sciences, 7*(12), 250.

Morton, B. (2016). Hong Kong's mangrove biodiversity and its conservation within the context of a southern Chinese megalopolis. A review and a proposal for Lai Chi Wo to be designated as a World Heritage Site. *Regional Studies in Marine Science, 8*, 382–399.

Narain, V., & Vij, S. (2016). Where have all the commons gone? *Geoforum, 68*, 21–24.

Polletta, F., & Jasper, J. M. (2001). Collective identity and social movements. *Annual Review of Sociology., 27*, 283–305.

Roe, D., Seddon, N., & Elliott, J. (2019). Biodiversity loss is a development issue: A rapid review of evidence. IIED Issue Paper. IIED, London.

Vargas-Lundius, R., & Sutti, R. (2014). Investing in young rural people for sustainable and equitable development. International Fund for Agricultural Development

Chapter 4
Building Resilient Rural Communities: A Summary of the Case Study and Prospects Beyond Lai Chi Wo and Hong Kong

Abstract Building rural resilience into the Lai Chi Wo SES has increased the economic opportunities of the village and strengthened the natural resource management, while safeguarding the cultural identity and natural environment of the area. Revitalising a rural area allows the continuation of a community's cultural heritage and so must be done with sensitivity to the local context, balances must be struck between maintaining traditions and innovating. The experiences and lessons gained from the Lai Chi Wo Programme are being developed and applied to neighbouring villages and, potentially the wider region, demonstrating the utility of the approach for revitalising peri-urban areas. The Lai Chi Wo experience demonstrates that peri-urban areas contain some unique advantages, which can be leveraged to strengthen the peri-urban interface to the benefit of both rural and urban communities.

4.1 Rural Communities as Resilient Social-Ecological Systems: Insights from Lai Chi Wo

Rural resilience requires the recognition of the integrated relationship between humans and nature and that economic and social systems are interlinked. People and their actions are required to support functioning, healthy ecosystems (Folke et al., 2005). Building resilient rural communities requires the SES to be able to adapt to external changes while maintaining its natural and cultural assets. This involves balancing vulnerabilities and functions in the region's ecosystem, economic and social dimensions (Heijman et al., 2007, Schouten et al., 2009).

Resilience empowers communities to react and adapt to social and ecological changes in their lives. In particular, it allows a SES to maintain functionality and avoid collapse in the face of endogenous or exogenous changes. An important part of this is the understanding and relationship communities have with the landscape, particularly in the modern context. By implementing the resilience understanding and factors identified in Chap. 2, the revitalisation efforts at LCW have focused on establishing rural/urban interconnections, a collaborative culture, and networked institutions. Each of these approaches and the important roles played by the bridging organisation will be summarised below.

© University of Hong Kong 2021
J. M. Williams et al., *Revitalising Rural Communities*, SpringerBriefs on Case Studies of Sustainable Development, https://doi.org/10.1007/978-981-16-5824-2_4

4.1.1   Rural/Urban Interconnections

Multiple interconnections between LCW, other peri-urban areas and urban Hong Kong served to re-brand this landscape for civil society and the government. This contributes towards two long-term and interdependent goals: re-establishing effective management regimes for village commons and supporting rural-based/urban–rural entrepreneurship for social, cultural and ecological benefits. The Programme revived social capital within the village, safeguarded cultural identity and cultivated links between urban and rural communities alongside bolstering agricultural practices. These allowed the village to explore sustainable socio-economic models, which build stronger ties and in relevance with the wider community and, in return, helps to maintain its heritage.

The LCW SES struggled with depopulation under the influence of urbanisation and changing food production systems in the broader context. Depopulation led to the breakdown of interdependent relationships between human and nature as management regimes for the village's common natural resources fall apart. This in turn created challenges for the area's biodiversity, soil quality, water management and increased vulnerability to extreme weather events. There is an increasing awareness that commons embedded in villages are part of an essential supporting system for the well-being and sustainability of the wider society. As such, revitalisation efforts had to create a sustainable economy for the village that was able to connect with and take advantage of its proximity to urban Hong Kong. Leveraging LCW's peri-urban position, the revitalisation programme focused on connecting the village with urban Hong Kong through building linkages between the village SES and SESs at a greater urban scale.

The LCW Programme has created and supported multiple diversified economic models for the village community, all of which contribute to the social and/or environmental sustainability of peri-urban areas and the wider territory. Various income streams and governance platforms were explored and implemented to build in redundancy and resilience to changes in the wider SES. The processes involved include supporting and/or bringing in local entrepreneurs through incubation and start-up schemes to create linkages of a variety of nature between the LCW community, resources and the wider urban community. For example, through the creation of new food production and processing channels, which focused on unique products made from local produce, it was also able to bring people into the village to rebuild a local community with the Indigenous villagers who have returned.

Multiple institutions were also developed, with the example of the LoCoKitchen as a market-oriented institution established to facilitate the building of a more localised and sustainable food system for Hong Kong. The Programme also invested in the villagers and in building strong social capital, both to safeguard the village culture but also to develop a new community to manage and govern the village and its resources. In this way, echoing Li et al.'s (2019) perspective, the revitalisation of LCW can be said to have created a sustainable and resilient rural community.

4.1.2 Establishing a Culture of Collaboration

Adopting a collaborative approach, the LCW case demonstrates that common pitfalls associated with more traditional top-down and bottom-up approaches can be avoided or, at least, mitigated. Rather than being government driven, the rural community acts as a key stakeholder in the process. This also differed from bottom-up revitalisation approaches, which rely on/are led by the villagers. In the LCW case, the involvement of civil society, through the facilitation of PSL and partnering organisations, opened up the village to the wider community, reconnecting urban and rural areas and strengthening the interface between the two areas. Doing so contributes significantly to enhancing the resilience of both urban and rural communities as their capacities to respond to shocks and stresses becomes greatly improved through such interconnections and collaborations.

By making rural development part of the public debate, the wider community becomes reacquainted with rural areas, developing a new understanding and sense of appreciation. The wide range of stakeholders with different expertise and knowledge brought with them invaluable ideas and feasible models for revitalising the village, while protecting and maintaining the traditional character and culture of LCW. Collaboration between artists, various types of experts, the villagers of LCW and those from other rural areas sparked numerous ideas to safeguard the cultural capital and local knowledge possessed by the Indigenous population, while introducing and experimenting with innovative ways to revitalise the village. Therefore, the traditional trap of overly focusing on agriculture and its modernisation (Liu & Li, 2017) was avoided.

The involvement of all types of actors in the revitalisation process not only helped to diversify development for the village, but also to (re)establish socio-ecological relationships. Ecologists and hydrologists helped to monitor the area's natural environment and its changes in relation to social activities through exercises such as ongoing biodiversity monitoring, ensuring the integrity of the surrounding ecological system was supported and accounted for in social activities. This meant that a solid level of understanding of the interaction between social and ecological processes could be developed as well as how such interactions may shift as changes occurred. On the basis of such understanding, the Programme, maintaining its flexibility, could then adapt its approaches and strategies accordingly. Incorporating learning and adaptation into the revitalisation process is crucial in the building of rural resilience.

4.1.3 Building Networked Institutions

While the importance of building or strengthening horizontal and vertical linkages between rural villages and SESs at other levels and scales is recognised, the success of doing so is subject to a myriad of factors related to the interactions between these systems. Especially at the peri-urban interface, even when a new institution appears to have the villagers' support, it needs to be mindful of the wider institutional context. This was demonstrated with the proposed shared company to manage communal income and expenses. While the villagers initially supported the creation of such a company, the Hong Kong legal system does not have a suitable registration category for a not-for-profit company, instead it would have had to be registered as a charitable organisation. The reclassification of the institution to a charitable organisation concerned some of the Indigenous villagers who raised questions regarding the aim and functions of the institution.

Strengthening local institutions also served to bolster resilience for the village. While many innovations or modernisations, such as the introduction of the sharing kitchen and improvement of facilities, were met with enthusiasm by the villagers, the introduction of shared companies and social enterprises struggled to gain acceptance. HakkaHome@LCW, for example, has faced issues of legitimacy since its creation, with the villagers being mistrustful of the social enterprise due to insufficient consultation and articulation of its potential functions, especially with the villagers living overseas. This indicates the need for further understanding on ways to build new and effective institutions in the peri-urban interface.

While still work in progress, the institutions at the village have become better connected with institutions at other levels. Progress made at LCW has provided important evidence that the revitalisation of villages at Hong Kong's peri-urban areas is possible and can benefit wider society. The strategies and approaches adopted by the Programme had been commended by the government, triggering implementable policy actions such as the establishment of a new office, the Countryside Conservation Office (CCO), in the government for handling rural revitalisation and public funding has been set aside for such projects. In 2020–2021 the CCO awarded 15 grants for rural revitalisation projects, totalling over HKD 78.3 million (EDP, 2021). Given the Programme's status as a pioneer in the field in Hong Kong, lessons learnt from the LCW experience has helped to inform the work of the Office as they find their feet in supporting other revitalisation efforts in Hong Kong's peri-urban territories.

4.1.4 Bridging Organisation

With the involvement of multiple stakeholder groups, coordination and structure were maintained through the presence of a bridging organisation, PSL. PSL performed crucial roles in trust building and aligning goals to create a coherent vision for the revitalisation process. Previous efforts to revitalise the village had been unsuccessful due to the inability for all the involved stakeholders to agree on a vision for the village. The various meetings, forums and workshops PSL organised to facilitate dialogue as well as their efforts to build trust with the villagers and green groups was essential in building a collaborative environment. Even though the success of LCW was premised on several factors, the role of the bridging organisation, which provides leadership, unlocks resources, provides capacity and generates wider support for the Programme, was especially significant.

4.2 Scaling up and Replication

Rural systems at the urban/rural interface across the world are at risk, or have suffered, from collapse due to external shocks or larger socio-economic forces (Abel et al., 2006; Motesharrei et al., 2014; Tenza et al., 2017). Consequently, the quantity and quality of ecosystems and their resources in both peri-urban and urban areas have been declining as urban areas encroach into rural and lines become increasingly blurred (MEA, 2005). This is occurring in both the developed world, where factors such as the capitalisation and commercialisation of food production systems are at play (Wästfelt & Zhang, 2016) and developing countries, where rapid urbanisation and expansion is ostensibly the solution to endemic rural poverty (Kestemont et al., 2011). The ultimate results are usually the outmigration of rural populations and a decline in agricultural or other traditional livelihoods. Peri-urban areas also face the 'double jeopardy' of having to deal with both globalisation and urbanisation as well as being the spaces either 'left behind' by globalisation or that urbanisation has 'created' (Clark et al., 2013; Wästfelt & Zhang, 2016).

As an example, Japan's rural areas are facing a rapid population decline as young people leave rural villages for urban centres and never return, leaving behind an, increasingly shrinking, elderly population. This trend began in the 1950s thanks to rapid economic growth. The rural economies often offer limited jobs, such as those in forestry, farming or manual labour, and so is unable to generate jobs to incentivise the younger population to remain. Some areas are able to attract tourists, but often not in the scale to build an employment base sufficient in maintaining young people (Knight, 1994). The Japanese government has introduced a variety of policies to encourage regional revitalisation, including its 'Chiho sosei' policy in 2014. Despite numerous commendable projects, rural populations continue to decline and there are struggles with maintaining and harnessing the local culture and identity of rural areas (Kusakari et al., 2018, Rausch, 2010). As a result, rural resilience, as well as the resilience of the economy, has declined and biological and cultural diversity lost as many of the rural farmers and residents act as ecological and cultural stewards (Hisano et al., 2018).

Villager in China have also seen the mass outmigration of the younger generation from rural areas due to economic development and urbanisation, which has resulted in the explosive growth of 'left behind' children. With both parents migrating, these children are left to be cared for by family members and are significantly more at risk of health problems than those under parental care (Li et al., 2015). The social ('Hukou') system in China, whereby rural migrants do not receive social support or welfare in urban areas, rapid urbanisation and migration in pursuit of economic opportunities in urban areas have caused the continued decline in rural areas and impedes the advancement of rural livelihoods (Jingzhong et al., 2010). In response, the Chinese Central government has been advocating rural tourism as a strategy since 2013. While rural tourism does represent a means to increase rural income and connectivity, it risks the loss of rural socio-cultural characteristics and agriculture, competing with agriculture for resources such as water and labour. The increased 'openness' of peri-urban areas due to tourism also risks eroding these areas, in some instances in China, rural land and villagers have been replaced by retirement resorts for urban residents (Yang et al., 2019). The importance of safeguarding village culture and heritage during the LCW Programme may be able to inform rural development in China to prevent tourism or urban influences from dominating and eventually subsuming such areas.

Rural areas at the urban/rural interface are able to benefit urban areas by supplementing their food supply, provide ecosystem services, enhance cultural value and increase economic resilience (Wästfelt & Zhang, 2016). The LCW experience provides options for governments, NGOs and villages across the globe to undertake a more diversified approach in developing new opportunities. The LCW case represents an alternative to a purely agricultural model of revitalisation and supports studies highlighting the importance of local context and culture in the benefits that peri-urban areas can provide for their urban counterparts (Wästfelt & Zhang, 2016). While agriculture is often the centrepiece of rural economies, the LCW experience has contributed to evidence that solely revitalising agricultural production is often not economically viable in modern peri-urban SESs (Meyer 2014). A resilient community should not be overly reliant on one economic model, but a myriad of them, which are also connected with ecological, social and cultural assets in the area. The LCW Programme also illustrates that there is space for a third actor in the revitalisation debate, civil society can play an important role in aiding the government and local communities in revitalisation efforts. This opens the door for new actors and arrangements in the revitalisation initiatives as well as new modes of governance. As such, the LCW experience provides options for governments, NGOs and villages all over the world to undertake a more diversified approach in developing new opportunities for the revitalisation of these SESs.

The lessons, experience and knowledge gained from the LCW Programme have inspired extended efforts to replicate and scale-up its impact, which enhances LCW's resilience and nested-ness across peri-urban systems. While this book has examined the revitalisation process of one rural village in Hong Kong, LCW, in great depth, lessons learnt from this journey have significant potential relevance to other peri-urban areas, where SESs have suffered similar problems of outmigration and loss of viability of socio-ecological connections. In particular, the LCW case can inform rural revitalisation efforts locally, including revitalising a small adjacent village and experimenting with a village cluster concept amongst neighbouring villages in Hong Kong, as well as regionally through building a regional network of revitalised villages across the Asia–Pacific. These will be discussed in more detail below.

4.2.1 Participatory Revitalisation Action Plan for a Village Cluster

The LCW revitalisation experience demonstrated how multisector collaboration and rural–urban partnerships can generate substantial social capital and innovation to revive the near desolate village. It also raised the awareness and knowledge of the wider society in Hong Kong about rural sustainability, building a network of stakeholders with a shared interest about the rural environment. Building on this, the Participatory Revitalisation Action Plan for a Village Cluster is being explored. It proposes a participatory approach to form a cluster of revived villages consisting of LCW and its neighbouring Mui Tsz Lam, Kop Tong, Siu Tan, Ngau Shi Wu, So Lo Pun and Sam A Tsuen. The participatory approach will be delivered through a robust partnership model, which is participatory among public members who come from different sectors, levels and disciplines (Fig. 4.1).

The initiative aims to recreate a network village cluster to provide the shared tangible and intangible infrastructure to support rural based sustainable livelihoods and rural stewardship in Hong Kong. The engagement of multiple stakeholders will be very similar to the LCW programme. Organisational stakeholders, Indigenous villagers, young talents, professionals and grassroots/community organisations as well as those in the city would be involved to develop a shared vision for the village(s)

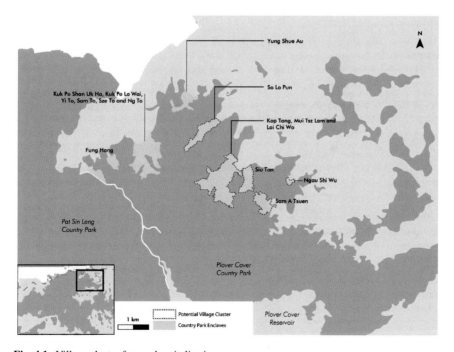

Fig. 4.1 Village cluster for rural revitalisation

and participate in the revitalisation initiatives. It also aims to establish a place based platform in the rural context for the attainment of ecological betterment and social wellness at the wider societal level, to facilitate the exchange of knowledge and skills as well as the matching of actors and resource for the incubation of rural revitalisation solutions and long-term rural–urban continuum in Hong Kong. Finally, it also aims to empower various communities of interest in the society to lead in sustainable place-based development.

The project will design, prototype, test and implement different conservation, education and socio-economic strategies for achieving sustainability in the peri-urban regions. This will be attained through three main scopes of activities: habitat and biodiversity management, place-making and identity building and tourism and education innovation. These activities will highlight the unique characteristics of the villages in the cluster so that they complement and support each other, creating a network for economic activity, cultural appreciation and environmental enjoyment.

The first part of the project involves a revitalisation project, funded by the CCO, at Mui Tsz Lam, a small village that is a 15 min walk from LCW. While the project draws on the knowledge from LCW, particularly in its aims to conserve architectural, ecological and cultural assets, it also branches out to focus more on citizen participation and creating 'change makers' to address rural issues. This will be done through a 'Citizen Scientist' and public engagement training as well as basic volunteer training, which will involve a range of engagement and education activities designed to involve and build the capacity of the general public in the restoration and protection of rural areas. Despite the shift in focus and activities from LCW's stakeholder driven approach, the LCW Programme does provide several important lessons about how to manage and govern such revitalisation projects.

The initial work at Mui Tsz Lam has demonstrated the importance of effective management and over-arching coordination. The introduction of the CCO has improved vertical coordination as the Office provides a direct interface with the different government departments, which helps to coordinate and communicate amongst them. The CCO is still establishing itself and so its role is not always clear, as a result, there can be confusion with communication or which channels issues should proceed through. This can create difficulties in resolving conflicts and co-ordination at the strategic planning level, especially as, being mainly comprised of architects and engineers, the CCO lacks a holistic perspective as well as the relevant expertise and knowledge regarding rural issues. The current teething issues with Mui Tsz Lam demonstrate the importance of the trust building phase at LCW and in establishing clear and open lines of communication between the villagers and external parties including the government. Being a new organisation, the CCO has yet to resolve these issues but it is committed to clarifying the terms of conditions of its funding awards and continues to maintain open communication to aid the resolution of issues.

4.2.2 APAC Initiative for Regional Impact

On an international level, the Programme team is developing an APAC Initiative for Regional Impact (AIRI) to scale up the impact of the knowledge and experience gained from the LCW Programme. AIRI will build a regional network of action-research institutions and action leaders to develop rural sustainability in the Asia Pacific region. The initiative builds on the successes of the LCW case and contributes towards attaining the 2030 Sustainable Development Agenda, notably the SDG 17 for strengthening global partnerships for sustainable development.

There are two major components of the AIRI. The first component is to build a regional consortium of action-research institutions in rural sustainability. This involves establishing a regional consortium through which participating action-research institutions will jointly promote sustainability. Such actions include the promotion of sustainability through collaborative research, collective actions at the regional scale and the empowerment of a community of 'Change Fellows' who are capable of creating innovative solutions to sustainability challenges. The Programme team is currently engaged in initial discussions with reputable action-research institutions in China and Thailand to develop this consortium.

The Programme team will provide knowledge and resources to support members of the consortium to function as intermediaries in their respective countries/regions. These intermediaries will then develop institutional infrastructure to support local competence and innovation, organise incubation activities and residencies, empower and connect leaders and practitioners in rural sustainability and build networks. Regular meetings will be held to share and develop knowledge as well as provide support. The intermediary institutions will also organise workshops and talent engagement events to disseminate knowledge and connect with local communities as well as identify local practitioners and leaders. These individuals will be invited to join the APAC Rural Sustainability Fellowship Scheme.

The aim is to recruit individuals from different countries and a range of sectors who are innovative, impact-driven, passionate and who have a track record of leading novel sustainability initiatives. The Fellowship scheme will recruit, connect, empower and recognise these leaders and exceptional practitioners to tackle sustainability challenges.

4.3 Concluding Remarks

While this book shares many lessons learnt from the LCW experience for revitalisation in a peri-urban village, it is important to note that the approach is not a panacea for every community at the urban–rural interface. The local context and unique cultural and societal characteristics must always be considered when implementing such revitalisation projects. Determining, likely through experimentation, the appropriate scale of a project/revitalised community is difficult but necessary.

Scaling up in the form of expanding revitalisation efforts on the same community risks blanket measures that may result in dampening the unique characteristics of the area, blurring the individual nature and character of rural areas.

Embarking on rural revitalisation provides the opportunity to continue that community's story and breathe into it new life. It requires making decisions on what to maintain and where to innovate, and how to strike a balance between the two in a way that safeguards the distinct characteristics of that SES. Revitalisation represents a new chapter for peri-urban SESs, but it is not the end point, the SES will continue to evolve, change and adapt, depending on the needs and decisions of the community, and in this way become truly resilient.

By strengthening the urban–rural connections of LCW with wider Hong Kong, the Programme has facilitated better management of the LCW SESs and addressed issues related to the interactions between the LCW SES and SESs at higher levels and scales. This allows for a two way flow of people and resources that can benefit both areas, not just the negative externalities from urban areas encroaching on the rural. As a result, increased integration and interaction between urban and rural communities has occurred. For example, where urban communities are working alongside rural in farming and developing products, while the Indigenous community share their cultural capital with the wider community and the modern world. The benefits of either area has been made more readily available to the other, alternative jobs and economic opportunities have been created for urban dwellers, while the villagers benefit from new ideas, innovations and manpower. In return, the villagers share their culture and traditional products with wider society, reconnecting individuals with their culture, providing niche local products and education and learning opportunities.

Through integrating several SESs the Programme has improved the sustainability of the rural community. The greater integration of LCW SES with SESs at other levels means that the system has greater adaptability, which was demonstrated at the farm level. Initially, the Programme sought to re-establish rice farming, a symbolic common for the modern LCW village. When it became evident that rice farming was no longer viable the Programme switched to growing coffee and established farming-agricultural sector-wider food system, a sectoral platform for coffee growing, processing, sale and appreciation. Initiatives such as these further develop the socio-economic viability of the village and local agricultural SESs through the creation of a new product at LCW and so new links with different sectors and industries, which creates a more developed and sustainable food system.

Many peri-urban SESs across the world are on the brink/have collapsed due to the impact of urbanisation and globalisation. While the cause for some is urban encroachment or a dominating influence of urban processes, for relatively remote peri-urban regions, desertion could be a common reason for dwindling commons as management regimes crumble. The LCW village was no exception, being a classic case of outmigration, loss of economic activity and near abandonment. Its revitalisation, however, demonstrates that peri-urban areas contain some important advantages, which can be leveraged to not only restore these SESs, but ensure their sustainability for the future.

The presence of resilient rural communities are not only beneficial for its members, but are valuable assets for the society at large. Therefore, the significance of understanding what makes up rural resilience lies not only for those interested in the rural or the peri-urban per se, but all who are concerned with sustainability of SESs at all levels and scales.

References

Abel, N., Cumming, D. H. M., & Anderies, J. M. (2006). Collapse and reorganisation in social-ecological systems: Questions, some ideas and policy implications. *Ecology and Society, 11*(1), 17.

Clark, J. K., Munroe, D. K., & Ramsey, D. (2013). The relational geography of peri-urban farmer adaptation. *Journal of Rural and Community Development, 8*(3), 15–28.

EDP. (2021). *CCFS approved projects.* The Hong Kong Government. https://www.epd.gov.hk/epd/english/environmentinhk/conservation/ccfs/ccfs_approved_projects.html Accessed June 25, 2021

Folke, C., Hahn, T., Olsson, P., & Norberg, J. (2005). Adaptive governance of social-ecological systems. *Annual Review of Environment and Resources, 30*(1), 441–473.

Heijman, W., Hagelaar, G., & Heide, M. (2007) Rural resilience as a new development concept. EAAE seminar Serbian Association of Agricultural Economists, Novi Sad, Serbia.

Hisano, S., Akitsu, M., & McGreevy, S. R. (2018). Revitalising rurality under the neoliberal transformation of agriculture: Experiences of re-agrarianisation in Japan. *Journal of Rural Studies, 61*, 290–301.

Jingzhong, Y., Yihuan, W., & Keyun, Z. (2010). Rural-urban migration and the plight of 'left behind children' in mid-west China. In N. Long, Y. Jingzhong, & W. Yihuan (Eds.), *Rural transformations and developments—China in context: The everyday lives of policies and people* (pp. 253–279). Edward Elgar.

Kestemont, B., Frendo, L., & Zaccai, E. (2011). Indicators of the impacts of development on environment: A comparison of Africa and Europe. *Ecological Indicators, 11*, 848–856.

Knight, J. (1994). Rural revitalization in Japan: Spirit of the village and taste of the country. *Asian Survey, 34*(7), 634–646.

Kusakari, Y., Chiu, D., Muasa, L., Takahashi, K., & Kudo, S. (2018). Entrusting the future of rural society through nurturing civic pride: Endeavors in Gojome Town Akita Prefecture of Japan. *Consilience, 20*, 104–114.

Li, Y., Westlund, H., & Liu, Y. (2019). Why some rural areas decline while some others not: An overview of rural evolution in the world. *Journal of Rural Studies, 68*, 135–143.

Li, Q., Liu, G., & Zang, W. (2015). The health of left-behind children in rural China. *China Economic Review, 36*, 367–376.

Liu, Y., & Li, Y. (2017). Revitalize the world's countryside. *Nature News, 548*(7667), 275.

MEA. (2005). *Ecosystem and human well-being: Biodiversity synthesis.* World Resources Institute.

Motesharrei, S., Rivas, J., & Kalnay, E. (2014). Human and nature dynamics (HANDY): Modelling inequality and use of resources in the collapse or sustainability of societies. *Ecological Economics, 101*, 90–102.

Okamuro, H., & Nishimura, J. (2020). What shapes local innovation policies? Empirical evidence from Japanese cities. *Administrative Science, 10*, 11.

Rausch, A. (2010). Cultural commodities in Japanese rural revitalisation: Tsugaru Nuri lacquerware and Tsugaru Shamisen. Brill

Schouten, M., van der Heide, M., & Heijam, W. (2009). *Resilience of social-ecological systems in European rural areas: Theory and prospects.* Paper prepared for presentation at the 113th EAAE

Seminar: The role of knowledge, innovation and human capital in multifunctional agriculture and territorial rural development, Belgrade, Republic of Serbia December 9–11, 2009

Tenza, A., Perez, I., Martinez-Fernandez, & Gimenez, A. (2017). Understanding the decline and resilience loss of a long-lived social-ecological system: insights from system dynamics. *Ecology and Society 22* (2), 15

Wästfelt, A., & Zhang, Q. (2016). Reclaiming localisation for revitalising agriculture: A case study of peri-urban agricultural change in Gothenburg, Sweden. *Journal of Rural Studies, 47*(A), 172–185

Yang, X., Hung, K., & Xiao, H. (2019). A dynamic view on tourism and rural development: A tale of two villages in Yunnan Province China. *Journal of China Tourism Research, 15*(2), 240–261.

Printed in the United States
by Baker & Taylor Publisher Services